Patient independent system to detect the electrical onset of temporal lobe epileptic seizure

V SRIDEVI

TABLE OF CONTENTS

1 INTRODUCTION 1

   1.1 The concept of seizure initialization 3
   1.2 Different types of epileptic seizures 3
   1.3 The need for seizure detection 6
   1.4 EEG measurement 8
   1.5 Seizure detection system 9
       1.5.1 Objectives of this research work 11
       1.5.2 Organization of the thesis 11

2 LITERATURE SURVEY 14

   2.1 Feature extraction 15
       2.1.1 Spike occurrence rate 15
       2.1.2 Synchronous neuronal activity 16
       2.1.3 Statistical measures 19
       2.1.4 Rhythmic oscillations 20
   2.2 Normalization Schemes 28
   2.3 Machine learning 30
   2.4 Thesis contribution 31
       2.4.1 The salient features 33

| | | 2.4.2 | Data flow of the proposed seizure detection system | 33 |

3 SCALP ELECTROENCEPHALOGRAM AND SEIZURES — 35

    3.1 Patient Dataset 35

        3.1.1 Data selection 37

        3.1.2 Channels Selection 38

        3.1.3 Sampling rate conversion 39

        3.1.4 EEG Filtering 42

        3.1.5 Electrical vs Clinical onset of seizures 43

        3.1.6 EEG Normalization 49

    3.2 Summary 54

4 DESIGN OF SEIZURE DETECTION SYSTEM USING NORMALIZATION SCHEME 1 — 55

    4.1 Feature extraction 55

        4.1.1 Signal Energy 56

        4.1.2 Entropy 56

        4.1.3 Statistical measures 59

        4.1.4 Spectral Energy 60

        4.1.5 Wavelet energy 60

        4.1.6 Spectral entropy 61

    4.2 Selection of optimal window length 62

    4.3 Significance of feature set 63

        4.3.1 Time domain features 64

        4.3.2 Frequency domain measures 74

        4.3.3 Significance of features - Quantitative analysis 76

    4.4 Reduced feature set 77

    4.5 Patient independent system 79

        4.5.1 Training of classifiers using signal energy features 82

        4.5.2 Training of classifiers using all three features 87

    4.6 Summary 96

5 DESIGN OF SEIZURE DETECTION SYSTEM USING NORMALIZATION SCHEME 2 — 97

	5.1	Optimal Features Set	97
	5.2	Automated Seizure detection	99
	5.3	Performance of classifiers on Cohort dataset	106
	5.4	Performance of LDA-SVM classifier	108
	5.5	Long EEG Recordings	111
	5.6	Testing the proposed algorithm on new recordings	116
	5.7	External Validity	120
	5.8	Utility in the Clinical study	123
	5.9	Summary	124
6	**DISCUSSION AND CONCLUSION**	**126**
	6.1	Major outcomes of this research work	126
		6.1.1 Sliding window length	127
		6.1.2 Significant features selection	129
		6.1.3 Patient dependent vs Patient independent system	131
		6.1.4 Selection of critical parameters	133
	6.2	Conclusions	135
	6.3	Future scope of this work	136
A	**LEARNING ALGORITHMS**	**138**
	A.1	Linear Discriminant Analysis (LDA)	138
	A.2	Naive Bayes classifier	141
	A.3	Decision tree classifier	141
	A.4	K Nearest Neighbour classifier	142
	A.5	Support vector machine	143

REFERENCES   **147**

CHAPTER 1

INTRODUCTION

Epilepsy is the most common neurological disease comprises a heterogeneous group of disorders, characterized by recurrent and unprovoked seizures due to the huge electrical discharges of large synchronized neurons. This hypersynchronous neuronal activity produces sustained action potential and repolarisation followed by hyperpolarisation, designated as Paroxysmal Depolarization Shift (PDS), as illustrated in Fig. 1.1. This abnormal activity changes the frequency and amplitude characteristics of the normal EEG and produces sustained oscillations with high amplitude and low-frequency spikes and slow-wave complexes. This huge potential affects the normal function of the neuronal network which leads to the loss of consciousness and involuntary body movements.

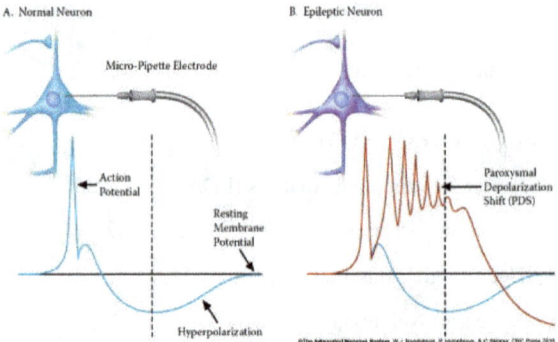

Figure 1.1: The action potential generated by normal and epileptic neuron is illustrated. The image used in this figure is adapted from The Integrated Nervous System, Second Edition (Fig. 11.7) by Walter J. Hendelman, Peter Humphreys and Christopher R. Skinner, 2017, CRC Press, Taylor and Francis Group, adapted with authors permission (Hendelman et al., 2017).

The cerebral cortex of the brain consists of many layers of neurons that are responsible for action potential propagation. There are two types of neurons named the Principal or Projection neurons and the Inter neurons or local neurons. The principal neurons are producing excitatory synaptic potential and the inter neurons produces inhibitory synaptic potential. In the neuronal network connection, both neurons are interconnected and the response of these neurons are summed up to activate the neighbour neurons. The inter neuron is connected with the projection neuron and provides feed-forward inhibition; The projection neuron is connected with inter neuron and then the inter neuron induces feed-back inhibition on projection neuron. Thus the combined activity of these two neurons control the synchronous activity of large neurons population. Moreover the inhibition and excitation of neurons are the electro-chemical processes. The neurotransmitters released by presynaptic neuron play the vital role in determining inhibition and excitation of action potential. The mechanism of neuronal excitability is referred as the hypersynchronous activity induced by increase in excitatory neurotransmitters, and/or decrease in inhibitory neurotransmitters. Apart from these, changes in voltage-gated ion channels or ligand gated ion channels and imbalance in ions concentration inside and outside the cellular space will influence neuronal excitability.

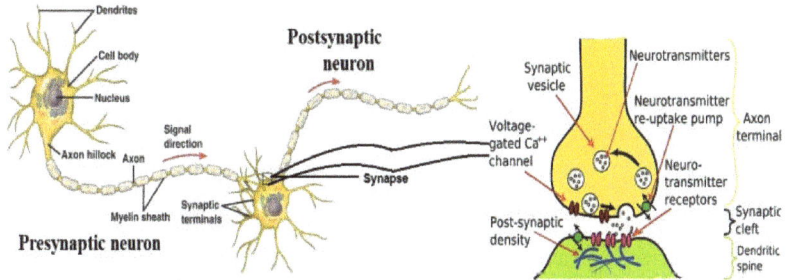

Figure 1.2: The neuronal network with pre-synaptic and post-synaptic neurons connection, sketch of synapse from (Julien et al., 2008).

The interconnection of presynaptic and postsynaptic neurons is shown in Fig. 1.2. The post synaptic neuron receives the signal from excitatory neurons and inhibitory neurons. The excitatory post synaptic potential (EPSP) triggers the depolarisation of the post synaptic neuron and inhibitory post synaptic potential (IPSP) produces the hyperpolarisation of post synaptic neuron. The post synaptic neuron receives excitatory and inhibitory signals from many neurons simultaneously within 1 to 2 milliseconds. Both the signals travel into the post synaptic neuron and reaches the axon terminal, are summed. If the summation result exceeds the resting potential threshold, the neuron will fire. Contradictorily, strong IPSP stops the firing of neuron. For an example, when we are in stress, the excitatory neurons sending the signals to large area of the brain and fires large population which leads to fear or feelings of panic. This mental state is modulated by the release of gamma-aminobutyric acid (GABA) by inhibitory neurons which produce the inhibition.

1.1 The concept of seizure initialization

Seizure may develop as a result of inheriting a mutation in a molecular mechanism that regulates neuron behavior, or organization of neuronal network. Alternatively, it may develop as a result of brain trauma such as a severe blow to the head, a stroke, a cerebral infection, or a brain malignancy. The symptomless gradual transformation from normal neuronal network to abnormal neural network is the period called as silent period. During the silent period, the changes occur in the organization of axons, loss of specific neuron (in particular inter-neurons and projection neurons driving inter-neurons), sprouting of axons to elicit large population of neurons and voltage-gated and ligand-gated channels property changes. This extensive change in synchronisation results permanent changes in the physical and biochemical structure of brain cells. The seizure is hyper-excitations of neuron, starts at a discrete region and spreads to all parts of the brain, either by stimulus or spontaneously. The recurrent hyper-excitation leads to muscle jerking corresponding to that brain area. When there is sufficient activation to fire the surrounding neuron, and the loss of inhibition of neighboring neurons leads to seizure propagation to neighboring areas.

1.2 Different types of epileptic seizures

During the course of seizure onset, the patients may have disturbed – mood, memory, consciousness, and involuntary movements. Depending on the brain region of seizure onset, seizure recurrence pattern, and the clinical characteristics, it is classified into many types. According to the new definition of seizure and epilepsy, the seizures are broadly classified into three types with respect to the cerebral region of onset and its spread (Falco-Walter *et al.*, 2018).

- Focal onset seizure
- Generalised onset seizure, and
- Unknown onset seizure

The classification of epileptic seizures is illustrated in Fig. 1.3. Focal seizures arise from a localized region of the brain's cortex, and have clinical manifestations reflect that region of the brain. It changes visual, auditory, olfactory, gustatory, vertiginous and psychic symptoms with or without motor activity. If the awareness impaired in any part of the focal seizure, it is named as focal impaired awareness seizure. As an example, a focal seizure originating in the temporal lobe, the part of the brain that processes emotions and short-term memory, may result in feelings such as euphoria, fear, and deja vu or hallucinations of taste or smell. Focal seizures may spread to involve other regions of the brain or the entire brain is called as focal to bilateral tonic-clonic seizures. As an example, a seizure originating in the left motor cortex may result in jerking movements of the right upper extremity. When the seizure spreads to adjacent areas, and the entire brain, whole-body convulsions resulted.

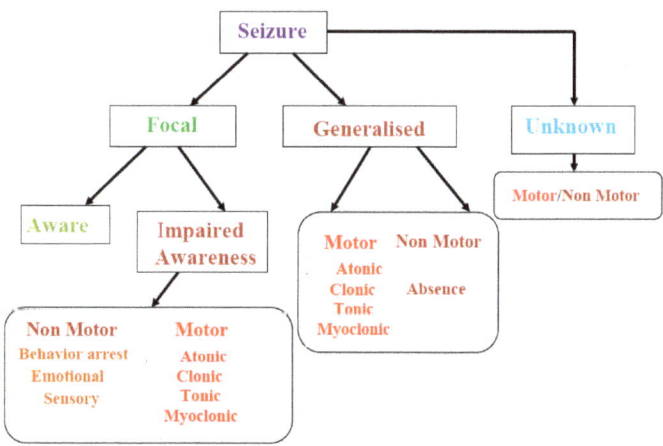

Figure 1.3: Types of epileptic seizures with respect to onset region and clinical symptoms are illustrated.

Generalized seizures begin with abnormal electrical activity at both hemispheres of the brain, and their corresponding EEG and behavior changes. The manifestations of such widespread abnormal electrical activity often includes the loss of consciousness. Motor manifestation of these seizures may include muscle stiffening which is called as tonic seizure. It affects the muscles on back, legs and arms and causes the individual to fall to the ground. Atonic seizure produces the sudden loss of muscle control which causes the individual to fall down. Myoclonic seizures produce sudden brief jerks or twitches of arms or legs of the individual. Clonic seizure associates with repeated

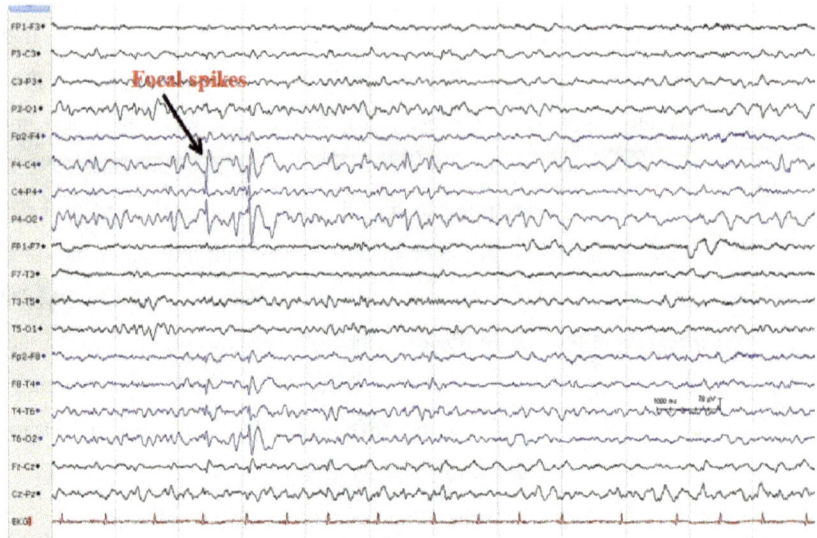

Figure 1.4: The right mid-temporal and central spikes consistent with the prototype of a focal epilepsy called Benign epilepsy with centro-temporal spikes (BECTS) is illustrated.

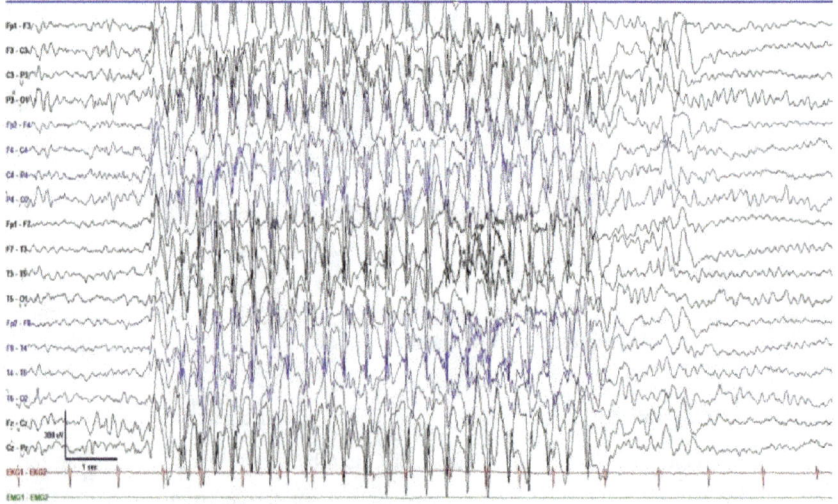

Figure 1.5: A burst of generalized frontally dominant 3 Hz spike and wave discharges, a prototype of idiopathic (genetic) generalized epilepsy called Childhood absence epilepsy is illustrated.

or rhythmic jerking muscle movements of face, neck and arms. A type of seizure produce rhythmic muscle jerking called as grand-mal or tonic-clonic seizure. Loss of consciousness, muscle stiffening and shaking, and tongue biting are the clinical characteristics of tonic-clonic seizure. The petit-mal or absence seizure is the most common in children, characterized by starring into space, or subtle body movements such as eye blinking and lip smacking. The third category, unknown seizure onset refers to the unknown onset with known motor and non-motor manifestations.

The focal seizures arise from the localized region have unique electrical characteristics that are more specific to the brain onset region. The Neurophysicians and Neurosurgeons highly rely on EEG patterns to identify, to quantify, and to localize the seizure onset (Fisher *et al.*, 2014). The EEG patterns with epileptiform discharges such as spikes, and sharp-waves are associated with the risk of occurrence of epileptic seizure. The spike is defined as a pointy shape potential clearly distinguished from the background EEG with asymmetric fall and rise, and a duration from 20 to under 70 milliseconds followed by a slow wave (waves with duration longer than alpha waves). The sharp wave is defined as a pointy peak with the duration from 70 to 200 milliseconds. The presence of spikes in the EEG, are associated with focal or whole area of the brain bilaterally. The focal spikes are associated with focal seizure with or without focal to bilateral generalization. Fig. 1.4 shows the right mid-temporal and central spikes consistent with the prototype of a focal epilepsy called Benign epilepsy with centro-temporal spikes (BECTS). Contradictorily, the generalized spikes associated with seizures that are non-focal at their onset (Fisher *et al.*, 2014). The scalp recorded seizure EEG consists of patterns with rhythmic oscillations evolving in delta, theta and alpha frequencies, rhythmic spiking – rhythmic recurrence of spike followed by slow waves. Generalized spike and wave patterns are the most common in absence seizures. Fig. 1.5 shows a burst of generalized frontally dominant 3 Hz spike and wave discharges, a prototype of idiopathic (genetic) generalized epilepsy is known as Childhood absence epilepsy.

1.3 The need for seizure detection

Worldwide, nearly 65 millions of people are affected by epilepsy. In India, epilepsy affects 12 million people and in one third of them, frequent, unpredictable seizures persist despite treatment by one or multiple anti-epileptic drugs. These types of seizures are known as medically intractable seizures. A major concern of patients with epilepsy, especially those with drug-resistant epilepsy, is the random and unexpected occurrence of epileptic seizures. Even brief episodes of staring or impairment of consciousness can lead to life threatening, especially if it occurs while the person is driving, swimming, climbing heights. The embarrassment of seizures often leads to social isolation and the loss of sense of well-being. Apart from that, in life the epileptic persons facing a lot of social problems in marriage, employment, and recreational activities.

A subset of patients with drug-resistant focal epilepsy are the candidates for epilepsy surgery. The success of resective epilepsy surgery depends upon the accurate localization of the origin of the seizures by long-term video EEG monitoring (VEM) and correlating the findings with structural abnormalities observable in magnetic resonance imaging (MRI). The ability to detect the electrical onset (the first alteration in the scalp-recorded EEG) of seizures may have a great impact on the management of epilepsy. Computerized seizure onset detection enable the new therapeutics and alerting systems that may ease the burden of intractable seizures. Moreover, a seizure warning system is made to trigger the injection of radioisotopes for use in an ictal single photon emission computed tomography (SPECT) study during the continuous EEG monitoring (Shoeb, 2009). A therapeutic system capable of detecting and responding to the electrical onset of a seizure could provide an opportunity to control the seizures by delivering fast-acting anti-epileptic drugs (AEDs) or electric stimulation such as vagus nerve or direct brain stimulation (Winterhalder *et al.*, 2003; Maiwald *et al.*, 2004; Bandarabadi *et al.*, 2015) that stops the progression of seizure prior to the clinical onset (the time at which the patient manifested the first observed alteration in behavior or motor activity). A warning system equipped with seizure onset detection could alert the patient of the seizure prior to the development of debilitating symptoms, or could notify the caretakers so that the consequences of a seizure can be limited. The warning that will be issued about the seizure onset may restore the individuals' confidence to overcome the limits of life that accompany seizures.

1.4 EEG measurement

The seizure onset and spread to other part of the brain is understood through the primary analysis of electroencephalogram (EEG). EEG is a multichannel recording of the electrical activity generated by the neurons of brain. When the EEG is measured by using noninvasive electrodes placed symmetrically on an individual's scalp, it is mentioned as scalp EEG; and when it is measured using electrodes placed on the surface of the brain cortex, it is mentioned as intracranial EEG. The popular non-invasive method of picking up the brain potentials using surface electrodes on the scalp follows 10-20 international electrode placement systems (Fig. 1.6). The onset of a focal seizure involves a change in activity on the few scalp EEG electrodes that lie above or near the site of the brain giving rise to a seizure; on the other hand, the onset of a generalized seizure involves activity on all scalp EEG electrodes.

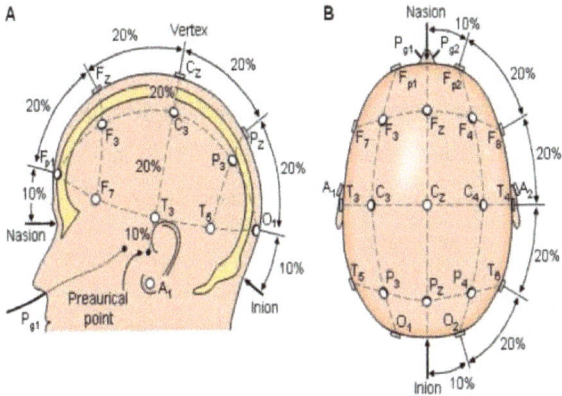

Figure 1.6: 10-20 electrode placement system to record scalp EEG. The image of the 10-20 electrode placement system used in this figure is adapted from Bioelectromagnetism, Principles and Applications of Bioelectric and Biomagnetic Fields (page no. 368) by J. Malmivuo, 1995, Oxford University Press, London, adapted with permission (Malmivuo and Plonsey, 1995).

In this method the active electrodes pickup the brain potentials with respect to the reference electrode and stored as instantaneous values. The electrodes are arranged, and the measurement are taken in two ways to visualize the EEG in required format: Longitudinal arrangement and Transverse arrangement. In Longitudinal arrangement, the electrodes are located from anterior to posterior head and the shape of the electrodes arrangement is in banana shape which covers the left and right hemisphere of the brain.

Similarly in Transverse arrangement, the electrodes are placed from left to right of the head which covers the brain activity from left to right parasagittal region and left to right temporal brain regions.

Depending on the depth of measurement, the electrodes are arranged closely or faraway, following the 10-5, 10-10 or 10-20 electrode placement systems. The measured raw EEG signal can be further modified into any desired representation for visualization of EEG activity of the brain. There are two major configuration with respect to the number of active electrodes involved to represent the EEG activity. The unipolar configuration compares each electrode potential with a common reference electrode potential, or the ensemble average of more than one electrodes potential. In general, the mid-line electrodes Fz, Cz, Pz or mastoid electrodes - A1 and A2 are used as reference electrodes in the unipolar configuration. The reference of ensemble average of all the electrodes potentials prevent the contamination of the signal by averaging all the electrodes potentials. Since the electrodes (Fp1, Fp2), (F7, F8), (Fz, Cz, Pz), and (O1, O2) are more contaminated by eye movement, eye-ball movement, mid-line vertex waves, and sweat artifacts respectively, are discarded for ensemble averaging the reference potential. Alternately, the bipolar configuration compares the two adjacent electrodes potential following double-banana montage to pickup the EEG activity from anterior to posterior brain region.

1.5 Seizure detection system

The electrical seizure detection is the identification of electrographic changes before the behavioral changes of the patients. The ability to detect the electrical onset of seizures can help with alternative treatment paradigms. Effective detection scheme also enables timely clinical interventions, reducing anti-epilepsy drugs usage and their various side-effects.

The EEG of epileptic patient is divided into three states: 1. Interictal phase refers to the normal state of the epileptic brain between seizures; 2. Preictal state refers to an altered brain state that is measurably different from a normal state; 3. Ictal state refers to the altered brain state induces the behavioral changes of the epileptic patient. The EEG patterns of interictal activity is illustrated in Fig. 1.7.

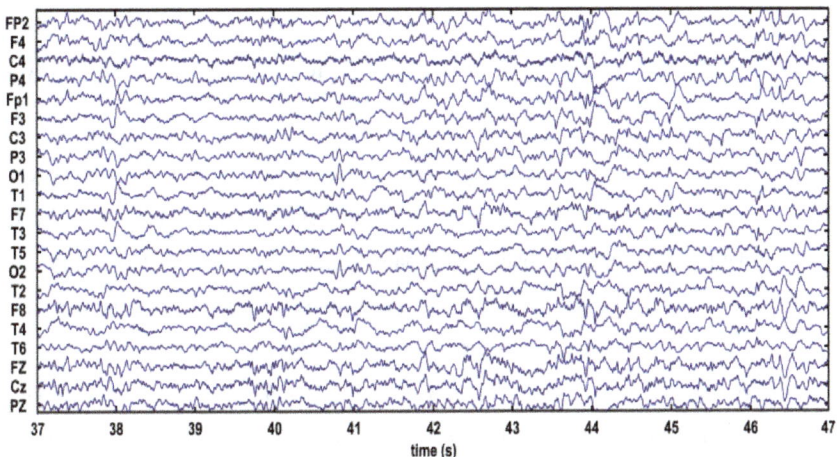

Figure 1.7: The interictal discharges within the scalp EEG of P5 patient.

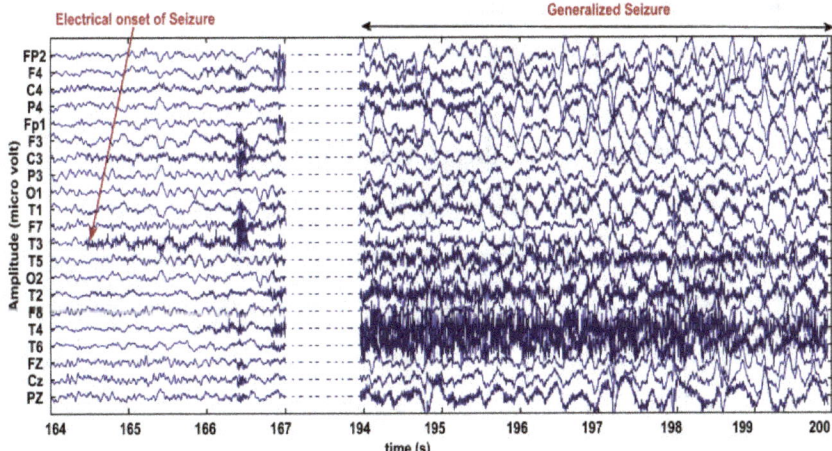

Figure 1.8: Example of a seizure within the scalp EEG of Patient P5. The seizure, which begins at 164.5 seconds, involves the appearance of a theta band rhythm on the channels (T3,T5) and developed as generalized seizure producing convulsive activity.

Due to hyper synchronous activity of neurons, the brain produces more rhythmic oscillations during the seizure. These changes result the chaotic behavior of the brain, from interictal to epileptic state evolves with different morphologies in different frequency bands, including delta (δ): 0.5 to 4 Hz, theta (θ): 4 to 8 Hz, alpha (α): 8 to 12 Hz, and beta (β): 12 to 30 Hz, showing amplitude depression and poly-spike patterns along with large amplitude waves. To prove the fact, the seizure patterns within the scalp EEG of P5 patient is illustrated in Fig. 1.8. The seizure, which begins at 164.5 seconds, involves the appearance of a theta band rhythm on the channels T3,T5 and developed as the generalized seizure producing convulsive activity after 194 seconds as shown in Fig. 1.8.

Long-term EEG recordings from patients, advancements in computer technology, digital EEG technology, and data storage allowed researchers to study the EEG with specifically employed mathematical tools in an attempt to identify changes or precursors to electrical or clinical seizure onset. During the last four decades, several studies have attempted to detect seizure occurrence based on changes in the EEG (Ramgopal *et al.*, 2014). Reliably detecting an epileptic seizure remains an unsolved problem even today and the clinical usefulness of those algorithms are very useful.

This research work motivated to analyze the impact of normalization scheme, sliding window length selection, features selection and machine learning algorithm for the design of patient independent system to detect the electrical onset of seizure.

1.5.1 Objectives of this research work

- To collect the real Electroencephalogram (EEG) data from hospitals.
- To examine the normalization effects on the features extraction from EEG.
- To analyze the sliding window length effects on the features extraction.
- To select the optimal features for the patient independent seizure detection.
- To train and validate the five classifiers for temporal lobe epileptic seizure detection.

1.5.2 Organization of the thesis

The Chapter 2 starts with the historical progress of epilepsy research starting from the first epileptic discharges recorded in living animals to the human epilepsy. Detailed reviews of the existing work on spike occurrence rate, synchronous epileptic discharges, and rhythmic oscillations based features for the development of seizure detection system are elaborated. Broad classification of machine learning algorithms used for the development of automated seizure detection system is given. The need for the design of patient independent seizure detection system which is the focus of this thesis is clearly discussed.

The Chapter 3 elaborates on the collection and pre-processing of scalp-recorded EEG used for the design of seizure detection system. The scalp EEG of 35 seizures recorded from 22 patients are utilized for this work. The selection of 5 minutes EEG from each variable length recording is explained. The filters used for removing supply line interference and its harmonics and, to obtain 1 to 200 Hz band limited EEG signal are explained. The choice of channels which provide more information on forthcoming seizures from 21 scalp electrodes, and the selection of optimal length of sliding window are demonstrated. Two different normalization schemes are explained as follows: Scheme 1 – Each seizure recording is normalized, using the minimum and maximum of that 5 minutes EEG; Scheme 2 – Each seizure recording is normalized, using the average minimum and maximum values computed from 29 seizure recordings of 18 patients used for training the model.

In Chapter 4, the features extracted from 5 minutes EEG normalized using normalization Scheme 1, and the design of classifiers for automated detection of the electrical onset of seizures are elaborated. The features – signal energy, approximate entropy, sample entropy, statistical measures, wavelet and spectral energies, spectral entropy – are extracted, and the selection of significant features are explained. The profile of spectral, and wavelet energy measures are compared in detail to select the simple, and efficient spectral energy measures for automated identification of normal and seizure condition. The characteristics of five classifiers – Linear Discriminant Analysis (LDA), Naive Bayes (NB), Decision Tree (DT), Support Vector Machine (SVM) and K-Nearest Neighbor (KNN) – are discussed. The five classifiers were trained using significant features – signal energy, spectral energy, spectral entropy – to detect the electrical onset of seizure. The performance of each classifier on patient independent data is evaluated, and the detailed comparison of performance metrics are given.

The Chapter 5 presents the patient independent system which minimizes the involvement of technicians to use the system on another patient. This section illustrates the way to extend the detector in Chapter 4 into a complete non-patient specific system to detect the electrical seizure onset. The most dominant features selected in Chapter 4 are extracted for each 5 minutes EEG normalized using normalization Scheme 2, and used to train five classifiers. Also, the performance of classifiers is evaluated by another test on a cohort containing more than one individual patient, and the results are reported. The SVM-based seizure detection system showed better detection capability in terms of sensitivity and specificity measures as compared to LDA, NB, DT and, KNN classifiers. Further improving the performance of SVM classifier, the features providing good separability of normal, and seizure condition among the four channels are selected by LDA classifier. The reduced features vector is used for SVM machine learning to automatically detect the onset of electrical seizure. The performance of LDA-SVM based system is compared with SVM based system, and the results are reported. Moreover the performance of SVM and LDA-SVM systems is compared on long EEG recordings of the 29 seizures of 18 patients, and the results are reported. Further, the proposed SVM based seizure detection system is tested on six seizure recordings of 4 patients which are not involved in the development of the system, and the classification performance is reported.

The Chapter 6 reviews the significant contribution of this thesis, and presents the Conclusion and Scope for further development of the system.

CHAPTER 2

LITERATURE SURVEY

Nearly one-third of the patients continue to have the seizure even after providing optimal anti-epileptic drug treatment. However, even the seizures are well controlled, living as an independent person is significantly lowered by anticipatory anxiety associated with the unpredictable seizures (Camfield and Camfield, 2010). The capability to detect the presence of clinical seizure, or to anticipate the onset prior to the seizure occurrence would provide greater advantages in epilepsy management. The seizure detection system must be capable of detecting the presence or absence of the seizure. A seizure detection algorithm consists of two major parts: feature extraction and classification. Selection of optimal features discriminating normal EEG, and epileptic seizure activity are essential for each detection algorithms. The choice of features varies based on morphological characteristics of inter-ictal spike discharges, synchronous neuronal activity, and continuous discharges of polymorphic patterns with different amplitude and frequencies. Depending on the choice of features, the characteristics of EEG are extracted in time domain, or from the transformed frequency space. The extracted characteristics are used to develop the machine learning algorithm for automated detection of seizure. Several classifiers using decision rules from simple thresholds to complex decision boundaries have been proposed to define the classification as a two-class or multi-class problem. For the past four decades, rigorous research on features extraction and machine learning based algorithms were being done by different groups of people. This chapter provides a detailed review on features extracted from time and frequency domain of the EEG, and the machine learning algorithms used for automated detection of electrical or clinical onset of seizure.

2.1 Feature extraction

2.1.1 Spike occurrence rate

The detection of the presence of interictal epileptiform discharges (IEDs) can help to confirm the clinical diagnosis of epilepsy. The characteristics, and localization of IEDs can help to identify the epileptogenic zone or to identify a particular type of epilepsy. In 1875, a British scientist Richard Caton, discovered the electrical properties of the brain, by recording spontaneous electrical activity from living brain of the animals. The human epilepsy research was started in the year 1924 and the first human EEG recorded by German physiologist and psychiatrist Hans Berger (Zifkin and Avanzini, 2009). Following the year, EEG plays the vital role in analysing the brain electrical activity for normal and epileptiform discharges. In 1934, Fisher and Lowenbach first demonstrated the presence of epileptiform spikes in animals EEG. In 1935, Gibbs, Davis, and Lennox described interictal epileptiform discharges in human EEG, and a year later in 1936 Gibbs and Jasper described the focal interictal spikes of absence seizure (Zifkin and Avanzini, 2009). Many studies investigated the temporal relationship between interictal spike rate, and the occurrence of seizure in focal and generalized epilepsy (Engel and Ackermann, 1980). F. Angeleri *et al.* (1972) observed varying trends (increases, decreases, and cyclic) in pre-ictal spike rate, and concluded that such variability precluded the prediction of upcoming seizures. Stevens *et al.* (1972) reported marked deviations (either increase or decrease) in spike rate 90 minutes prior to seizures but never indicated whether these changes are the consistent feature to detect the upcoming seizure. Lieb *et al.* (1978) and Gotman *et al.* (1982) observed no systematic changes in the rate of pre-ictal spike activity recorded with depth electrodes. In contrary to these results, Lange *et al.* (1983) showed systematic pre-ictal changes involved spike activity in the medial temporal lobes in patients whose seizure onset to be initiated unilaterally from the temporal lobe.

These analyzes concluded that the epileptic focal patterns of spike activity tended to decline significantly in rate of occurrence, several minutes prior to seizures. The later study utilized prolonged EEG recordings of 6 patients whose medication levels were stable; spike rates were sometimes found to increase seconds prior to the seizure, and concluded that the repeated seizures caused an increase in spike rate (Lange *et al.*,

1983). Gotman and Marciani (1985) reported that the level of spiking activity did not influence the probability of seizure occurrence. They stated that a high spiking rate did not increase the likelihood of seizures and a low rate did not decrease it. Later, Selvitelli *et al.* (2010) confirmed that there is no association between the presence or absence of interictal epileptiform discharges (IEDs) or its occurrence rate, and the most recently determined seizure frequency. Also stated that the higher IED incidence was seen in subjects with longer epilepsy duration (p = 0.04).

However, prior studies have not found a consistent association between the presence or frequency of IEDs, and clinical epilepsy severity, possibly because of EEG recording techniques, and differences in subject characteristics. Even though there is no consistent association between the spike rate and seizure occurrence, several studies proposed morphological filter based (Nishida *et al.*, 1999; Xu *et al.*, 2007), neural network based (Gabor and Seyal, 1992; Ozdamar and Kalayci, 1998; Ko and Chung, 2000), clustering based (Inan and Kuntalp, 2007; Shen *et al.*, 2009), wavelet based (Indiradevi *et al.*, 2008), eigen value based (Fukami *et al.*, 2018) techniques aimed to identify the spike discharges in the ongoing EEG for the localization of epileptogenic foci.

2.1.2 Synchronous neuronal activity

Before 1970s, it was believed that the seizure was an abrupt process which could not be predicted by any means, and later in 1970s, the rigorous work on seizure detection/prediction started with the finding of Vigilion and his co-workers, the seed of seizure may develop over several hours before its onset (Litt and Echauz, 2002; Carney *et al.*, 2011). Initially, the linear methods were used to detect the seizure precursors based on amplitude and frequency variations. In 1980s, after the development of non-linear systems theory, the non-linear techniques were used to extract the features to detect/forecast the seizure onset. The univariate measures pickup the EEG variations from single electrode recording which reflects the dynamics of that brain region where as the bivariate measures pickup the signal from more than one recording region to study the interaction among the brain areas.

Due to hyper synchronous activity of neurons, the brain produces more rhythmic oscillations during seizure. Most previous seizure detection algorithms attempted to identify the signatures of seizure activity within fixed length sample epochs. An Artificial Neural Network (ANN) based system quantifying the amplitude, slope, curvature, rhythmicity, and the frequency components of scalp recorded EEG was developed to detect the clinical seizure. The filtered 1 to 27 Hz EEG was separated into 2 seconds non-overlapping windows to extract a set of 31 context parameters. The mean, standard deviation (SD), and coefficients of variations were calculated for parameters (Webber *et al.*, 1996). In the same year, Learning Vector Quantization (LVQ) neural network based algorithm was proposed to characterize the three types of temporal EEG samples such as definite epileptic discharges (EDs), propable EDs, and non-EDs collected from five patients admitted in National Institute of Mental Health and Neurosciences, Bangalore. The 12 seconds, 0.05 to 32 Hz band limited EEG segments containing eye blink related artifacts were incorporated into the training data to ensure the reduction of common confounding patterns among the various defined classes of EEG (Pradhan *et al.*, 1996). The Monitor algorithm, and its later incarnation named Sensa, analyzed 2 seconds epochs of EEG by first applying a digital notch filter and then decomposing EEG into half-waves, using rules to identify significant amplitude peaks. If the coefficient of variation for an epoch, computed as the ratio of the variance to the square of the mean of the half-waves, is less than 0.36, and their frequency is between 3 and 20 Hz, then the epoch is identified as potentially part of a seizure. An algorithm called, CNet, a non commercial seizure detection algorithm, discriminates seizures from non-seizure activities using a detector based on a self-organizing map. The detector operates on two-dimensional Fast Fourier Transforms (2DFFT) of EEG data. The EEG are filtered to obtain 4 to 18 Hz signal, and then FFT is computed for every 125 millisecond for frequency ranges 0 to 32 Hz. For each frequency bin, a second FFT is then applied to produce the 2DFFT (Gabor, 1998). An algorithm called Reveal, by using matching pursuit, small neural network-rules, and connected-object hierarchical clustering, had sensitivities and false detection rates that compared favorably with two other algorithms (Sensa and CNet) (Wilson *et al.*, 2004). Sixteen different features were evaluated in their potential ability to detect seizures from scalp EEG recordings containing temporal lobe (TL) seizures. Features include spectral measures, Minimums and maximums, Zero crossings, Omega complexity, phase synchronization and the Brain Symmetry Index (BSI). Among them, BSI is the most relevant individual feature to detect electroencephalographic seizure activity in TLE with unilateral epileptiform discharges (Van Putten *et al.*, 2005). The BSI was revised as a function of spatial (sBSI) and temporal (tBSI) changes, and as a function of the number of channels to localize the hemisphere of seizure onset (Van Putten, 2007).

Another automated seizure detection system named IdentEvent was developed for efficient scanning of long-term scalp EEG recordings and the identification of ictal EEG segments. The system processed multi-channel scalp EEG signals and translated them into three main EEG descriptors: pattern-match regularity statistic (PMRS), local maximum frequency (LMF), and amplitude variation (AV). The PMRS quantifies the degree of signal regularity of a time series by estimating the likelihood of a signal pattern repeated within the time series. PMRS values become smaller when the signal is in a less complex state. Therefore, during the seizure, due to the highly organized and ordered patterns of ictal EEG, PMRS values drop significantly from their background interictal values (Kelly *et al.*, 2010). Another seizure detection system was developed by utilizing PMRS as primary EEG descriptor along with local maximal frequency (LFmax), amplitude variation (AV), local minimal, maximal amplitude variation (LAVmin and LAVmax), and maximal amplitude in a high frequency band (AHFmax) EEG descriptors. The spatio-temporal patterns of these descriptors were used for detecting ictal EEG epochs as well as rejecting the EEGs with significant artifacts. The false detection rate of this algorithm is significantly smaller than that of Reveal algorithm at different level of detection sensitivities (Shiau *et al.*, 2010).

The interictal EEG in the epileptogenic region is more synchronised than the other area of the brain. These phathologically increased synchronization of chaotic brain from interictal to epileptic state has been identified by using entropy based measures. Pincus (1991), Richman and Moorman (2000) developed the entropy measures to detect the deterministic nature of chaotic time series. In the following year, Kannathal *et al.* (2005) proposed different types of entropy such as Shannon's entropy, Renyi's entropy, Kolmogorov–Sinai entropy, and approximate entropy (ApEn) to investigate the EEG collected from a group of five normal, and epileptic patients. A neural network based system trained using ApEn features was developed to examine the EEG to detect the normal and epileptic condition (Srinivasan *et al.*, 2007). The entropy measures such as wavelet entropy, sample entropy (SampEn), and spectral entropy were used to train two different neural network models namely Recurrent Elman network (REN) and radial basis network (RBN) to classify the normal, interictal and ictal EEG samples

(Kumar *et al.*, 2010). In an another work using the same data-set, four entropy features namely ApEn, SampEn, Phase Entropy 1, and 2 (S1 and S2), were fed to seven different classifiers: Fuzzy Sugeno Classifier (FSC), Support Vector Machine (SVM), K-Nearest Neighbour (KNN), Probabilistic Neural Network (PNN), Decision Tree (DT), Gaussian Mixture Model (GMM), and Naive Bayes Classifier (NBC). The Fuzzy classifier was able to differentiate the three classes with good accuracy (Acharya *et al.*, 2012). Another study using the same dataset, an optimized sample entropy (O-SampEn) algorithm with fast computational speed, was proposed and combined with extreme learning machine (ELM) to identify the EEG signals regarding the existence of seizure or not (Song *et al.*, 2012). Alternate to sample entropy, Fuzzy entropy measures trained SVM, showed better performance to differentiate the seizure and non-seizure samples (Xiang *et al.*, 2015). In 2017, an another automatic technique was developed to detect the epileptic activity in EEG signals using multi-domain features and nonlinear analysis to improve the performance of detection system. Then, multiple features such as intrinsic mode functions (IMFs), spectrum entropy and approximate entropy (ApEn) are extracted from five frequency bands of clinical interest in order to increase the classification robustness and accuracy of the epileptic seizure detection. Furthermore, a dimensionality reduction algorithm of the principal component analysis (PCA) along with the feature ranking technique of analysis of variance (ANOVA) was applied to eliminate irrelevant or redundant features. This measure was used to analyze the EEG of healthy person and seizure and seizure free EEG of epileptic patients (Wang *et al.*, 2017).

2.1.3 Statistical measures

The characteristics of EEG vary during the normal, interictal, and ictal states of the brain. These changes are captured by using statistical measures namely mean, median, mode, root mean squared amplitude, standard deviation, variance, skewness, and kurtosis. These features are directly applied to the raw EEG or in the transformed space, to measure the central tendency, and dispersion of sample population. In the previous studies, the statistical features were used to detect or forecast the seizure onset. In 2003, McSherry et al. applied the profile of variance to EEG recordings of 19 seizures collected from a group of 11 patients, each having medial temporal lobe epilepsy.

The profile of variance increases during or prior to the seizure onset. Also, using surrogate data demonstrated that the increases in variance simply reflect decreases in the non-linear measure of correlation density (McSharry et al., 2003). These statistical measures have shown significant changes during the preictal period compared to the interictal state (Rogowski et al., 1981; Mormann et al., 2005; Rasekhi et al., 2013). More specifically, a decrease in variance coupled with an increase in kurtosis has been reported during the preictal period (Mormann et al., 2005). A new statistical method combined with LDA classifier was proposed to discriminate the epileptic EEG signals from normal signals. Among many possible metrics for statistical distances, opted Mahalanobis distance to measure the statistical difference of power spectrum in each frequency band. From the distance measures, significant frequency bands which showed the largest distance between the normal and epileptic power spectrum were selected to train the LDA classifier for seizure detection (Choe et al., 2010). In the same year, the variance, skewness, and fluctuation intensity of wavelet coefficients at 2,3,4, and 5 scales of detail, and approximation of 0.3 to 70 Hz EEG were extracted (Minasyan et al., 2010). The statistical features like mean, variance, skewness and kurtosis were included in the features set training the SVM for automated neonatal seizure detection (Temko et al., 2011). In contrast, these measures were compared with wavelet, spectral and time domain measures and reported as the least significant measures to discriminate the normal, and seizure samples (Logesparan et al., 2012). In another work, standard deviation, variance, and root mean square measures along with spectral measures were selected as significant features for the detection of generalized epileptic EEG signals (Ravish et al., 2013).

2.1.4 Rhythmic oscillations

The normal brain produces rhythmic oscillations specific to the functions. For example, alpha (α) rhythms in 8 to 12 Hz band recorded at rest with eyes closed state. The beta (β) rhythms at 12 to 30 Hz, and gamma (γ) rhythms in 30 to 50 Hz band exits during mental and motor tasks. The theta (θ) frequency oscillations at 4 to 8 Hz band exits during memory tasks or sleep. Similarly, the delta (δ) waves in 0.5 to 4 Hz band characterizes slow wave sleep. The EEG activity in these traditional bands are the main focus of EEG research, and intensively studied the normal and diseased states of brain.

The epilepsy condition alters the synchrony of neuronal network which leads to rhythmic oscillations or rhythmic spike discharges. In 1982 Gotman (Gotman, 1982), proposed an automated system to detect the rhythmic epileptic discharges in the EEG. The coefficient of variation was used as a measure of 'rhythmicity' of the EEG. The coefficient of variation is the ratio of the standard deviation to the mean; it was small when all the half-waves were of similar duration and large when there was great variability in the duration of the half-waves of one epoch lasting 2 seconds. It measures only the regularity of their duration, independent of the frequency of the waves. The seizure detection was occured when the relative amplitude was at least equal to 3, and the sustained paroxysmal rhythmic activity with a fundamental frequency anywhere between 3 to 20 Hz. In 1995, an automated seizure warning system for 24 seizure types of 17 patients was developed. The seizure templates extracted from surface, and depth EEG recordings were used to train the classifier for automated seizure detection (Qu and Gotman, 1995). Later, several spectral and wavelet based features have been proposed to capture these rhythmic oscillations during the seizure state. The 3 level Daubechies 4 wavelet was used to decompose the 5 to 45 Hz invasively recorded EEG from mesial temporal lobe epileptic patients. The power spectral density (PSD) of signal in different level was used to recognize the seizure onset, and for the short time prediction of clinical seizure (Osorio *et al.*, 1998).

The increased rhythmicity gradually increase the energy level of the EEG signal. The energy variations of the EEG has been measured to study the profile for inter-ictal, pre-ictal, ictal, and post-ictal phases. The accumulated energy variations were studied on five patients continuous EEG, and it was found that the energy increased 50 minutes before the seizure onset (Litt *et al.*, 2001). Adeli *et al.* (2003), extracted the discrete wavelet transform based features to detect the 3 Hz spike and slow wave epileptic discharges in absence seizures. In another study, the wavelet energy based analysis was used on depth EEG recordings to separate the inter-ictal, and ictal activities (Gigola *et al.*, 2004). In the same year, wavelet transform based features were used to train the SVM classifier for detecting different types of 139 seizures recorded from pediatric patients. In another research work, Daubechies 4 wavelet was used to decompose the EEG into six levels, and the corresponding frequency range of 0.5 to 25 Hz which captures seizure onsets of various electrographic manifestations (Shoeb *et al.*, 2004). In another study, a 5-level Daubechies-4 wavelet transform was proposed to detect the seizure in scalp recorded EEG. The scalp recorded EEG was decomposed into 50 to 100 Hz, 25 to 50 Hz, 12 to 25 Hz, 6 to 12 Hz, 3 to 6 Hz, and 0 to 3 Hz bands signal. The 0 to 3 Hz band was discarded due to the occurrences of nonictal sleep EEG that can be frequent in this band. The information in 50 to 100 and 25 to 50 Hz bands was used to estimate the amount of EMG artifact present in the EEG, and the information in 12 to 25 Hz was used to characterize the alpha activity. The 3 characterizing measures designed by Khan and Gotman (2003), namely relative average amplitude, relative scale energy, and the coefficient of variation of amplitude were derived directly from the wavelet coefficients for seizure detection (Saab and Gotman, 2005). Mormann *et al.* (2005) observed the contrast result that the signal energy decreases 30 minutes before the seizure onset. In the same year, Esteller *et al.* (2005) reported that the pre-ictal signal energy variations are not reproducible and this should be combined with other features to improve the prediction performance. In 2006, power spectral density (PSD) in delta, theta, alpha, and beta frequencies were extracted for the development of association rules to classify epileptiform transients (ET), muscular artifacts, eye blink artifacts, and sharp alpha activity (Exarchos *et al.*, 2006). In another study, DWT based features were used to classify the EEG recorded from normal and epileptic patients with absence seizures. The recorded EEG was decomposed into 5 levels namely A5 (0 to 3.125 Hz), D5 (3.125 to 6.25 Hz), D4 (6.25 to 12.5 Hz), and D3 (12.5 to 25 Hz), D2(25 to 50 Hz), and D1(50 to 100 Hz) using Daubechies-4 wavelet filter. Among them, four frequency bands, A5, D5, D4, D3 which lie in the δ, θ, α, and β frequency bands respectively were used to train the fuzzy inference system (Subasi, 2006; Subasi *et al.*, 2019). Alternately, the intracranial EEG recorded from healthy subjects and epileptic patients were used to develop the seizure detection system. The power spectral density of selected channels were extracted using Welch FFT method, and trained the decision tree classifier for distinguishing the normal and epileptic EEG (Polat and Gunes, 2007).

However, in previous studies, the different polymorphic ictal EEG patterns, had not been taken into account in the seizure detection algorithm. The six most common ictal morphologies (delta, theta, alpha, and beta - rhythmic activity, amplitude depression, and polyspikes) of 91 seizures (37 focal, 54 secondarily generalized) recorded from 57 patients were analysed. The seizure morphology plays a crucial role in increasing the detection performance of the generic seizure detection system (Meier *et al.*, 2008). In 2009, shobe developed a scalp EEG based seizure detection system, utilizing the spectral features of 8 uniform bandwidth filter bank spans 0 to 25 Hz frequency range using each filter that are 3 Hz wide (Shoeb, 2009). In another study, the rhythmicity of ictal intracranial EEG of partial seizure had been captured by using dominant frequency, average amplitude, and coefficient of variation of amplitude. The dominant frequency is the frequency which has the maximum spectral power. The average amplitude was computed by taking the average of amplitudes of the peaks within 3 to 30 Hz frequency band. The coefficient variation was measured as the ratio of mean and the standard deviation of the peak-to-peak amplitudes of waves in EEG (Aarabi et al., 2009). In the same year, Temko et al. (2009) developed a system utilizing the statical, spectral, and wavelet measures to detect the neonatal seizures. The total power in 0 to 12 Hz band, power in frequency bands of width 2 Hz from 0 to 12 Hz, spectral edge frequency, dominant-peak frequency, the energy in the 5th coefficient of Daubechy 4 wavelet decomposition that corresponds to 1 to 2 Hz were extracted to capture the rhythmic oscillations of synchronised neurons. In 2010, a set of 58 features including spectral, wavelet features were extracted to capture the rhythmic ictal and muscle activity of 86 seizures recorded from 25 patients with focal epilepsy. The relative power in 0 to 4 Hz, 4 to 8 Hz, 8 to 12 Hz, 12 to 16 Hz, 16 to 20 Hz, 20 to 24 Hz, 24 to 28 Hz), 28 to 32 Hz, and 32 to 70 Hz, mean frequency, peak frequency and bandwidth of the peak frequency were extracted using Fourier Transform (FFT). The EEG signal is decomposed into five levels using Daubechies 4 mother wavelet to cover the most EEG rhythms and waveforms (Minasyan et al., 2010).

In the same year, probabilistic neural network (PNN) based seizure detection system trained using discrete wavelet energy, entropy and standard deviation features extracted from 100 to 200 Hz, 50 to 100 Hz, 25 to 50 Hz, 12.5 to 25 Hz, 6.25 to 12.5 Hz, 3.125 to 6.25 Hz and 0 to 3.125 bands was reported. The EEG signal was decomposed into six levels using Daubechies 4 mother wavelet and the wavelet energy was found to be the best feature and achieved the accuracy of 99.33% in classifying the epileptic and non-epileptic EEG signals (Gandhi et al., 2010). Further improving the accuracy and decreasing the execution time of seizure detection system Gandhi et al. (2012) proposed a algorithm based on discrete wavelet packet transform (DWPT) with energy, entropy, standard deviation, mean, kurtosis, skewness and entropy estimation at each level of the wavelet tree. Among the total number of 384 features, only 45 optimal features were selected through Discrete harmony search with modified differential (DHS-MD) operator feature selection algorithm. The PNN classifier 10-fold cross validated using these 45 optimal features achieved 100% classification accuracy.

In 2011, the modified set of spectral features extracted by Shoeb *et al.* (2004), were extracted from intracranial EEG for the detection of seizure onset. The intracranial (iEEG) based seizure detection system utilizing the spectral features of 8 uniform bandwidth filter bank spans the 0.5 to 35 Hz frequency range using each filter that are 3 Hz wide whereas the range of 35 to 105 Hz range is covered by a lower density of filters, each with a bandwidth of 15 Hz (Kharbouch *et al.*, 2011). In another study by Iscan *et al.* (2011), the power spectral densities in 0.5 to 4 Hz, 4 to 8 Hz, 8 to 13 Hz, 13 to 25 Hz, and 25 to 40 Hz were selected as significant features to train the KNN and SVM classifiers to distinguish the healthy and epileptic EEG samples. The spectral power features extracted from raw, time-differential, and space-differential EEG were used to differentiate pre-ictal and inter-ictal EEG signals. It was reported that the cost-sensitive SVM trained using spectral power extracted from time/space-differential Electrocortocogram (ECoG) obtained 86.25% sensitivity and 0.1281 false positives rate in out-of-sample testing (Park *et al.*, 2011). Yoo *et al.* (2012) designed a multichannel-based processor called system on chip (SoC) for detecting the seizure activity. The device has 8 data acquisition channels, feature extraction module and classification engine. The EEG signal was divided into small segments and the energy was calculated from these segments, which results in a more discriminative feature vector. The SVM classifier trained using signal energy featutres was used to detect the rapid-eye blink patterns as this is similar to the generalized seizure and has more energy as compared to non-seizure patterns. Bhople (2012) proposed a fast Fourier transform (FFT) based seizure detection algorithm. The FFT based features extracted from Bonn EEG dataset, fed into the multilayer Perceptron (MLP) and generalized feed-forward neural network (GFFNN) achieved 100% accuracy. In the following year, Teager energy (TE) based approach to discriminate EEG signals corresponding to nonseizure (eyes open, eyes closed, or interictal) and seizure (ictal) intervals was proposed. Unlike the instantaneous signal energy which is only proportional to squared instantaneous amplitude, TE is proportional to the squared product of both instantaneous amplitude and instantaneous frequency. This new energy measure is therefore capable of responding rapidly to changes in both amplitude and frequency (Kamath, 2013). Khamis *et al.* (2013) proposed a patient specific seizure detection algorithm based on frequency-moment signatures. At the sampling rate of 256 Hz, signatures were calculated over 32 seconds from 8192 pairs of data points consisting of electrodes difference T6-P4 from right hemisphere (RH) and T5-P3 from left hemisphere (LH). The discrete Fourier transform of the 1024 point subsets (with frequency spacing of 0.25 Hz) with 50% overlapping window were calculated, resulting in a total of 15 subset transforms from each RH block and LH block. The spectral powers at a given frequency from the RH and LH were combined into a single quantity. The signature elements were found by subtracting normalized central moments of the subset distribution from the mean, to measure the consistency of the spectral power at a given frequency over all subsets. The seizure measure was the logarithm of the probability that a signature belonged to a control set of non-seizure signatures. Bandarabadi *et al.* (2014) proposed a robust and low complexity threshold based algorithm for seizure detection. The relative monopolar spectral power (ratio of spectral power within 9 to 12 Hz and spectral power within 0.5 to 3 Hz) and relative bipolar spectral power (ratio of spectral power within 12 to 18 Hz to 0.5 to 3 Hz) extracted from intracranial Electroencephalogram (iEEG) exceeding the threshold values was used for seizure onset detection (Bandarabadi *et al.*, 2014, 2015). The spectral power in delta, theta, alpha, beta and gamma frequency bands were calculated and normalized by the total spectral power. In addition to that the spectral edge power is described as minimum frequency up to which 50% of the overall power of 0 to 40 Hz band is contained. The spectral edge power, spectral edge energy and normalized spectral power of sub bands appeared to be the best features for identifying the onset of seizures (Rasekhi *et al.*, 2015).

The performance of neonatal seizure detection system was validated and clinically tested on 0.5 to 13 Hz pediatric EEG (Temko *et al.*, 2009, 2011, 2015). In 2016, a pediatric non-patient specific seizure detection system utilizing stationary wavelet transform (SWT) based features was developed. The SWT is also known as time invariant orthonormal wavelet representation which can be obtained by modifying the basic scheme of discrete wavelet transform (DWT). The time invariance in SWT algorithm is achieved by removing the down-sampling stages of DWT, and replacing them by up-sampling the filter coefficients at each stage. Thus, the output of each level of decomposition has the same number of samples as the original signal. Each 2 seconds windowed EEG was decomposed into 6 levels using Daubechies 4 mother wavelet and

SWT was calculated for each segment. The frequency bands of (D3) 15.41 to 33.09 Hz, (D4) 7.69 to 18.56 Hz, (D5) 3.84 to 8.28 Hz, (D6) 1.91 to 4.16 Hz approximately equal to beta, alpha, theta, and delta brain rhythms were considered for features extraction. Then, the Mean and Peak Frequencies (MF and PF) were calculated on the power spectral density (PSD) of D3, D4, D5, and D6 bands. The extracted features were used to train LDA and neural network (NN) classifiers for automated seizure detection (Orosco et al., 2016). Swami et al. (2016) extracted energy, standard deviation, root-mean-square, Shannon entropy, mean values and maximum peaks features from dual-tree complex wavelet transform (DTCWT) coefficients of six level decomposed EEG signal. These features sets were used to train the general regression neural network (GRNN) for the automated classification of ictal and non-ictal samples of Bonn and SGRH, Delhi datasets. Further analysing the same algorithm using energy and standard deviation features on EEG signal recorded at the rate of 200 Hz at Neurology and Sleep Centre (NSC), New Delhi, it was reported that the spectral components within the range of 0 to 25 Hz improve the accuracy of epileptic and non-epileptic samples classification (Swami et al., 2019). Hamad et al. (2017) used discrete wavelet transform to extract the coefficients D1, D2, D3, D4 and A4 corresponding to 30 to 60 Hz, 15 to 30 Hz, 8 to 15 Hz, 4 to 8 Hz and 0 to 4 Hz respectively using four level Daubechies 4 wavelet. The features selection, and the SVM classifier parameters were optimized using Grey Wolf optimizer (GWO) for automated detection of seizure. In 2017, the increasing rhythmic signal amplitude compared to average EEG spectrum was used to trigger the seizure detection (Furbass et al., 2017). In another study, the signal energy features optimized using genetic algorithm were used for implementing the seizure detection algorithm. A simple effective way of detecting a given seizure EEG is to compare the energy of a small window (called foreground) with the energy of a much larger window (background). If the foreground's energy is significantly larger than the background's, then it is likely that the foreground window is part of a seizure (Baldominos and Ramon-Lozano, 2017). Hu et al. (2019) proposed a Convolutional Neural Network (CNN) and SVM based preictal, ictal, and interictal states classification algorithm utilizing mean amplitude spectrum (MAS) features. The amplitude spectrum of 0.3 Hz to 70 Hz EEG signal from 18 channels were calculated and divided into 19 frequency sub-bands. The MAS on each of the 19 frequency sub-bands is then computed for each channel to form a MAS map of size 18×19. Finally, the MAS map is fed to a CNN for feature extraction and SVM was used for the epileptic state classification.

Few studies analysed the efficiency of multiple time and frequency domain features in training the classifiers for automated detection of epileptic seizures. A set of three time domain and two frequency domain attributes namely average amplitude, average duration, coefficient of variation, dominant frequency, and average power spectrum, extracted from each epoch lasting 2.56 seconds were applied to the adaptive structure neural network (ASNN) and the back propagation (BP) algorithms for automated epileptic seizure detection (Weng and Khorasani, 1996). Sixteen features including spectral symmetry in 1 to 25 Hz band, symmetry of phase synchronization in 0 to 4 Hz, 1 to 4 Hz, 4 to 8 HZ and 8 to 13 Hz bands, number of zero crossings, number of minima and maxima, Global field amplitude and frequency, mean amplitude and Omega complexity were extracted. The classification algorithm trained using the combination of measure of symmetry characteristics, mean amplitude and number of manima and maxima features significantly improved the performance of temporal lobe epileptic seizure detector (Van Putten *et al.*, 2005). Minasyan *et al.* (2010) used 58 features including relative power in delta (0 to 4 Hz), theta (4 to 8 Hz), alpha (8 to 12 Hz), sigma (12 to 16 Hz), beta1 (16 to 20 Hz), beta2 (20 to 24 Hz), beta3 (24 to 28 Hz) and beta4 (28 to 32 Hz), and in the high-frequency band (32 to 70 Hz), mean frequency, peak frequency, and bandwidth of the peak frequency, average duration of half-waves, coefficients of variation of half wave duration and amplitude, and non-linear energy. In addition to that variance, skewness, and fluctuation intensity of wavelet coefficients at 2 to 5 scales of detail, and level 5 of approximation, the complexity measures such as spectral entropy, wavelet subband entropy, and autoregressive goodness of fit were used to train the recurrent neural network (RNN) detecting the onset of seizures. Logesparan *et al.* (2012) performed a study identifying the optimal features that gives best discrimination between epileptic seizures and the background EEG in the minimally pre-processed scalp EEG data. The core features were categorized into, 17 time domain, 8 Discrete Wavelet Transform (DWT)-based, 4 Continuous Wavelet Transform (CWT)-based and 6 Fourier transform (FT)-based features. There were 35 core features present, and then the DWT-based features and 2 FT-based features (power and spectral entropy) were calculated, based upon using different frequency bands. The 0 to 25 Hz frequency band have been selected to capture the epileptiform activity, which at the same time reduces signal disruptions due to high frequency artifacts. The DWT and FT based features were calculated across the frequency bands result from a standard five level decomposition of the EEG signal – D3(12.5 to 25 Hz), D4(6.25 to 12.5 Hz), D5(3.125 to 6.25 Hz), and A5(0.16 to 3.125 Hz) (Logesparan et al., 2013). A set of 52 features including power, 1st difference, 2nd difference, Hjorth features, Higher Order Crossings (HOC), spectral band power in 1 to 4 Hz, 4 to 8 Hz, 8 to 10 Hz, 8 to 12 Hz, 12 to 30 Hz and 30 to 64 Hz frequency bands, and Root Mean Square (RMS) and Recursive Energy Efficiency (REE) features extracted from DB4 wavelet six level decomposed EEG signal recorded at 512 Hz sampling rate were used to train the classifiers for automated seizure detection (Hernández et al.).

2.2 Normalization Schemes

The EEG normalization plays vital role in generalizing the seizure detection system across different people and age group. While majority of the studies concentrated more on the features extraction and features selection to emphasize the ongoing seizure activity in the EEG, very little attention has been given to the EEG normalization techniques that have great impact on the features highlighting the seizure activity. The normalization is performed in two ways: (i) Normalizing the raw EEG (in time domain) (ii) Normalizing the features (in feature space) without impair the seizure detection performance. The normalization techniques may vary in terms of the mathematical function (such as the mean or median), amount of background data used to calculate the normalizing factors, and placement wherein the signal processing chain the normalization is applied (Logesparan et al., 2011). For the design of deep neural network based seizure detection system the features were extracted from the EEG normalized to zero mean and unit variance (Zhao et al., 2020). Many of the existing seizure detection systems trained using the features normalized in features space. The SVM based system proposed by Meier et al. (2008), each feature sample is normalized separately by replacing it with a measure for its change over the recent past. Each feature sample was compared with its recent history (past 5 seconds), and its longer-term history (past 25 seconds). These two time series were compared by a floating rank sum test (Wilcoxon), yielding the probability P values. This transformation results in a mapping of the feature values space into a P values space for the entire feature set, normalized according to the probability of observing a change within the previously measured time window. In another study, the wavelet coefficients were normalized by using an envelope detector (Casson and Rodriguez-Villegas, 2009).

In a SVM based seizure detection system proposed by Rasekhi et al. (2013), a set of 22 linear features were normalized using three normalization methods – divide by max, divide by mean of the absolute values and divide by the standard deviation of the feature samples. It was reported that the feature samples divided by the maximum value of that feature vector performed better in discriminating the seizure and non-seizure condition. In another study, SVM based neonatal seizure detection has been improved applying feature baseline correction (FBC) normalization scheme on 103 features vector. For each channel feature vector, the FBC was done by calculating the average non-seizure feature value and selecting the optimal threshold that best separates the seizure and non-seizure condition (Bogaarts et al., 2014). The EEG baseline value may vary due to circadian rhythm, changes in the state of the patient's vigilance, a response to medication, or the changes in EEG recording quality (Hartmann et al., 2011; Duun-Henriksen et al., 2012). For optimal FBC functioning, the non-seizure EEG baseline segment needs to be updated to adapt to the changes in background EEG. In another study done by Logesparan et al. (2015), the impact of five EEG normalization techniques namely median decaying memory, mean memory, standard deviation memory, peak detector and signal range on discriminating the seizure EEG from non-seizure EEG activity were analysed. It was reported that the median decaying memory (MDM) to be the best normalization approach for using line length features to discriminate between seizure and non-seizure EEG. The MDM technique computes the feature normalization factor, based on the on-going unsupervised update of the baseline EEG buffer. However, during long-term continuous EEG the presence of artifacts, periodic discharges, sharp spikes, or a burst-suppression pattern can corrupt the baseline EEG update that may hamper the seizure detection performance. To overcome this limitation, novelty-MDM technique proposed by Bogaarts et al. (2016) rejects these EEG patterns (novel epochs) which are different from the baseline EEG for computing the normalization factor. It was reported that the novelty-MDM technique performed better than FBC and MDM normalization techniques. These normalized features were used for training the classifiers for the automated detection of seizure and non-seizure epochs.

2.3 Machine learning

The machine learning algorithms are one of the significant parts of automated detection system. There are two major types of machine learning algorithms: data based, and rule based approach. The former one using the actual data set to build the model, and as models are exposed to new data, they are able to adapt independently without human intervention. The system learns from the previous known (training) data samples to produce reliable, and repeatable decisions on samples which are unknown. Alternately, the later one using the set of rules derived from knowledge obtained from the data set. Moreover, the data based learning algorithms are further classified into supervised and unsupervised learning. The supervised learning algorithms are trained using the labeled data samples whereas the unsupervised algorithms using the data samples without any target information. The research on epilepsy used simple models with threshold technique to more complex boundaries for the automated discrimination of seizure and non-seizure patterns. In 1996, Gotman used the threshold technique to detect the seizure. For each epoch, the channels detection was made if: the relative half-wave amplitude exceeded 3, the average half-wave duration was between 25 and 150 milliseconds, and the coefficient of variation of half-wave duration was less than 0.6 (Gotman *et al.*, 1982; Gotman, 1999). The multimodal algorithm utilizing EEG, EMG and ECG signal used threshold to trigger the seizure detection (Furbass *et al.*, 2017).

The selection of threshold for fixing the boundary for normal EEG and epileptic patterns significantly changes the sensitivity and specificity of the system. Most of the studies used supervised learning to define the boundary between seizure and non-seizure to improve the performance without expert intervention. Several classifiers using simple to complex decision boundaries have been proposed to define the classification as a two-class or multi-class problem (Saab and Gotman, 2005; Iscan *et al.*, 2011; Quintero-Rincon *et al.*, 2016; Kumar *et al.*, 2010). The Linear Discriminant Analysis (LDA) classifier trained using time domain, frequency domain, and complexity measures (Greene *et al.*, 2008), statistical spectral features (Choe *et al.*, 2010), Approximate entropy, sample entropy derived from DWT (Wang *et al.*, 2017) distinguished the normal and seizure patterns. The Quadratic Discriminant Analysis (QDA) classifier was configured to improve the performance of seizure detection using

Fuzzy entropy (FuzzyEn), and distribution entropy (DistEn) (Li *et al.*, 2018), and time domain measures (Chua *et al.*, 2011). The parametric based learning – Bayes classifier was trained using discrete wavelet transform and entropy based measures for seizure detection (Acharya *et al.*, 2012; Wang *et al.*, 2017; Saab and Gotman, 2005). The Support Vector Machine (SVM), a statistical machine learning algorithm, has been used in a wide range of biomedical applications, including automated seizure detection (Meier *et al.*, 2008; Shoeb and Guttag, 2010; Kharbouch *et al.*, 2011; Temko *et al.*, 2011; Acharya *et al.*, 2012; Sitt *et al.*, 2014; Xiang *et al.*, 2015; Kafashan *et al.*, 2017; Wang *et al.*, 2017; Hamad *et al.*, 2017; Subasi *et al.*, 2019). Furthermore, the artificial neural network (ANN) has been trained using various features for the seizure detection system (Pradhan *et al.*, 1996; Webber *et al.*, 1996; Srinivasan *et al.*, 2007; Acharya *et al.*, 2012; Kumar *et al.*, 2010; Zhou *et al.*, 2018; Emami *et al.*, 2019). Several studies used unsupervised learning techniques such as Kohonen self-organising net, C-means clustering for automated seizure detection (Gabor, 1998; Inan and Kuntalp, 2007). Apart from, few other studies used rule based algorithms such as neural network rules, association rules, adaptive neuro fuzzy inference system (ANFIS), dynamic fuzzy neural network and decision tree (DT), to detect the seizure onset (Wilson *et al.*, 2004; Kannathal *et al.*, 2005; Exarchos *et al.*, 2006; Subasi, 2006; Polat and Gunes, 2007; Acharya *et al.*, 2012). The regression algorithms using wavelet entropies and Tunable Q wavelet transform (TQWT) have been proposed for the development of seizure detection system (Wang *et al.*, 2017; Sharaf *et al.*, 2018).

2.4 Thesis contribution

In the last four decades, several studies have been attempted to detect the seizure occurrence, based on changes in the EEG (Ramgopal *et al.*, 2014; Jory *et al.*, 2016; Ulate-Campos *et al.*, 2016; Sharmila and Geethanjali, 2019). However, reliably detecting an epileptic seizure remains an unsolved problem even today, for the following reasons: First, because electrical and clinical features of epileptic seizures differ from patient to patient and even between seizures in the same patient, it is difficult to develop a generic algorithm to forecast the clinical seizures reliably (Minasyan *et al.*, 2010; Furbass *et al.*, 2017; Emami *et al.*, 2019). The majority of studies that have investigated patient-specific detection paradigms have utilized extensive machine training strategies for the application to individual patients (Saab and Gotman, 2005; Kharbouch *et al.*, 2011; Shoeb *et al.*, 2004; Khamis *et al.*, 2009; Minasyan *et al.*, 2010; Chua *et al.*, 2011; Selvakumari *et al.*, 2019). These patient specific systems need extensive training for classifiers to reconfigure the system for individual patient. Also, the majority of studies developed the algorithm to detect the generalized or focal seizure from temporal and extra temporal region which needs knowledge on the selection of electrodes positions to pick-up the seizure activity (Aarabi *et al.*, 2009; Furbass *et al.*, 2017; Selvakumari *et al.*, 2019). Few studies have not specified the number of channels used to detect seizure activity (Hunyadi *et al.*, 2012). Few studies used all the electrode recordings make the detection more complex and time consuming (Exarchos *et al.*, 2006; Wang *et al.*, 2017; Emami *et al.*, 2019).

Second, the methods used to detect the seizures have differed widely, and included a number of univariate and bivariate measures comprising both linear and nonlinear approaches with different classification algorithms, thereby making inter-study comparisons difficult.(Greene *et al.*, 2008; Minasyan *et al.*, 2010; Temko *et al.*, 2011). Third, the majority of studies have used intracranially recorded EEG data because they are less prone to artifacts than scalp EEG signals (Aarabi *et al.*, 2009; Kharbouch *et al.*, 2011; Geng *et al.*, 2016). As intracranial EEG is invasive and can be undertaken only in a minority of patients, it would not be appropriate to generalize the findings to the majority of patients who undergo only scalp-recorded EEG (Le Van Quyen *et al.*, 2001). Finally, the influence that AED withdrawal can have on the electrical characteristics of the seizures in an individual patient is largely unknown.

In order to weed it out difficulties as mentioned above, this study is carried out to explore the utility of scalp-recorded EEG in detecting the electrical onset of seizures in a uniform cohort of well-characterized patients with drug-resistant temporal lobe epilepsy (TLE). To achieve this objective, applied a set of time and frequency measures – signal energy, spectral energy, and spectral entropy – to identify changes in the EEG at the electrical onset of seizures. An essential feature space is chosen to discriminate the normal, and seizure patterns for the development of an automated system for the detection of the electrical onset of seizure. The performances of five different classifiers, namely the Linear Discriminant Analysis (LDA), Naive Bayes (NB), Decision Tree (DT), Support Vector Machine (SVM) and K Nearest Neighbor (KNN), are examined for the clinical utility of the system. It was envisaged that the proposed algorithm would identify the electrical onset of seizures and would allow short-term prediction of clinical seizure onset, which might provide a window for diagnostic and therapeutic opportunities in the management of people with epileptic seizures.

2.4.1 The salient features

- The proposed patient non-specific seizure detection system does not require prior knowledge on channel selection, and data selection for training the classifier. This improves the comfort level of end users, the customers (Neurologists and Neurotechnologists).

- The patient independent normalization scheme is introduced in this research work. By using common normalization factors, the algorithm can be able to compare the EEG amplitude across the patients.

- The performance of classifiers are validated by using the cohort patients data. The proposed SVM shows good sensitivity and specificity compared to LDA, NB, DT and KNN classifiers. The algorithm is found suitable for patient non-specific seizure detection system.

2.4.2 Data flow of the proposed seizure detection system

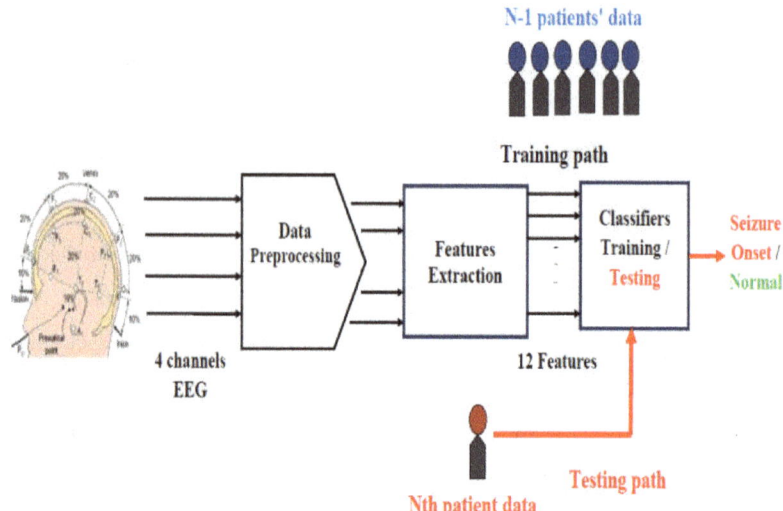

Figure 2.1: Steps in EEG data collection and processing for the automated seizure detection system.

The scalp recorded EEG is pre-processed, and time and frequency domain features are extracted. The features showing large variations at electrical onset of seizure are selected as significant features to discriminate the normal and seizure samples. The significant features of N-1 patients recordings are used to train the five classifiers, and Nth patients' recording is used for testing the classifier. Hence the system ensure the development of patient independent system for temporal lobe epileptic seizures. The proposed seizure detection system is tested on the unknown seizure recordings which are not used in the development of the system. The following chapters provide the detailed explanation of each part of the system illustrated in Fig. 2.1.

CHAPTER 3

SCALP ELECTROENCEPHALOGRAM AND SEIZURES

The popular non-invasive scalp recorded EEG picks up the state of brain with different activities such as awake and sleep. In particular, the dominant frequency and spatial distribution of EEG activity during the awake state is different from the state of sleep. Based on the frequency, an EEG signal is classified as delta (δ) wave if its dominant frequency component is below 4 Hz, a theta wave (θ) if frequencies are from 4 to 8 Hz, an alpha wave (α) when frequencies are from 8 to 12 Hz, a beta wave (β) when frequencies are from 12 to 30 Hz, a gamma wave (γ) when frequencies are above 30 Hz. Within the scalp EEG, seizures manifest as a sudden redistribution of energy depending on the spatial location of seizure onset (Shoeb, 2009). Furthermore, the EEG electrodes demonstrating these spectral energy changes also vary among the patients and the site of seizure origin.

Moreover the risk free scalp EEG is preferred for long term EEG monitoring. In this research work, scalp recorded EEG is utilized to analyse the seizure precursors. This chapter presents the patients dataset and its preprocessing to extract the features of EEG used for automated seizure detection.

3.1 Patient Dataset

The data collected from two hospitals were used for this research: dataset_1, consists of 11 patients (3 male; 8 female), collected at Sree Chitra Tirunal Institute of Medical Sciences and Technology, Trivandrum, Kerala, India and dataset_2 consists of 11 patients (6 male; 5 female), collected at Fortis Malar Hospital, Chennai, India. Among them, 21 patients had unilateral seizure onset (12 Right; 10 Left) and one patient with bilateral seizure onset rising from right and left hemispheres of the brain. The clinical, EEG and MRI findings of these patients were consistent with the diagnosis of temporal lobe epilepsy (TLE). The video- scalp EEG monitoring (VEM) was undertaken as per standard protocol, utilizing 21 scalp electrodes placed according to the standard 10-20 EEG electrode placement system which has been described in literature (Orosco *et al.*, 2016). The electrical onset (the time at which the first alteration in the scalp-recorded EEG) and clinical onset (the time at which the patient manifested the first observed alteration in behavior or motor activity in the synchronized video) of all 35 seizures recorded from 22 patients were visually identified by three Neurologists. Among the 22 patients, 18 patients data consisting of 29 seizure recordings were used for developing the automated seizure detection algorithm. The ages of the 18 patients (6 male, 12 female) ranged from 16 to 46 years. The 6 seizures recorded from the rest of 4 patients were used for testing the performance of the automated electrical seizure onset detection system. Permission was obtained from the Institute Ethical Committee to utilize the scalp-recorded VEM data for this research study. A summary of the clinical characteristics is provided in Table 3.1.

Table 3.1: Summary of clinical data

Patient ID	Sex	Age	Onset Hemisphere	No. of seizures
P1	M	43	Bilateral	3
P2	M	39	Right	3
P3	F	46	Left	1
P4	F	44	Left	1
P5	F	30	Left	1
P7	F	28	Left	3
P8	F	31	Right	2
P11	F	33	Left	3
P12	F	23	Left	1
P13	F	29	Right	1
P14	M	37	Right	1
P15	F	21	Right	1
P16	F	19	Right	1
P17	F	28	Right	2
P18	M	42	Left	1
P19	M	44	Right	1
P20	F	39	Right	2
P21	M	16	Right	1
P22	M	-	Left	2
P23	M	-	Left	2
P24	F	-	Right	1
P25	M	-	Right	1

Note: The electrical onset of seizure episodes in patients P6, P9 and P10 were not clearly visible in the EEG. Hence these recordings were excluded from the analysis.

3.1.1 Data selection

Previous studies are helpful to analyze the different preictal durations of 5 minutes to 240 minutes to predict the electrical onset of seizure, and the mean prediction time varied from a few seconds to minutes with varied accuracies (Maiwald et al., 2004; Mormann et al., 2005; Bandarabadi et al., 2015; Rasekhi et al., 2015). Using the algorithm named Advanced Seizure Prediction Via Pre-Ictal Relabeling (ASPPR), the best intervention time proposed for the prediction of the electrographic onset of seizures was 1 minute (Moghim and Corne, 2014). Hence in this work, for consistency across all 29 seizure recordings of 18 patients, 3 minutes EEG prior to clinical onset is selected. The localization of the epileptogenic focus by ictal Single Photon Emission Computed Tomography (SPECT) is most reliable if the isotope is injected within 30 seconds of the ictal EEG onset (Rathore et al., 2011).

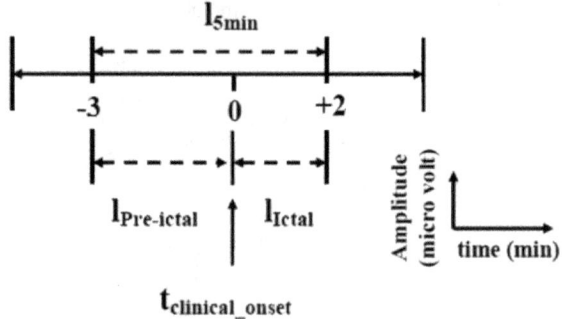

Figure 3.1: General schema of 5 minutes EEG data selection with respect to clinical onset of seizure.

Considering the seizure activity, a seizure usually lasts for 30 seconds to 2 minutes followed by a postictal phase lasting several minutes (Engel, 2001; Rasekhi et al., 2015). In this work, based on the length of available EEG, the data is split into preictal and ictal phases. In this work, the 3 minutes long EEG sample immediately before the clinical onset is defined as *preictal* phase, whereas the *ictal* phase is defined as 2 minutes EEG recording clipped immediately after the clinical onset as shown in Fig. 3.1. Among the 29 seizure recordings, five seizure recordings – P7_s2, P7_s3, P8_s1, P8_s3 and P12_s1 had less than 2 minutes of ictal EEG. In these five seizure recordings, the EEG severely contaminated by artifacts during the clinical seizure was discarded.

3.1.2 Channels Selection

Each recorded EEG consists of 21 scalp electrodes from temporal (T1,T2,T3,T4,T5,T6), frontal (Fp1,Fp2,F3,F4,F7,F8), central (C3,C4), parietal (P3,P4), occipital (O1,O2), and midline (Fz,Cz,Pz) regions of the brain. The raw EEG was converted into common average representation (Mormann *et al.*, 2005; Cherian *et al.*, 2012; Indiradevi *et al.*, 2008) in which, at each time instant the reference voltage is obtained by ensemble averaging F3,F4,C3,C4,P3,P4,T1,T2,T3,T4,T5 and T6 electrode potentials. For reference voltage calculation, frontopolar (Fp1,Fp2), lateral frontal (F7,F8), occipital (O1,O2) and midline (Fz,Cz and Pz) electrodes were discarded. The 10-20 electrode placement system of 21 electrodes is shown in Fig. 3.2. The different set of electrodes specific to the region (temporal, frontal, central, parietal and occipital) pickup the focal or location related epilepsy. The Temporal Lobe Epilepsy (TLE) is the most common form of focal epilepsy resistant to medications. There are two types of temporal epilepsy as one arising in the hippocampus, the parahippocampal gyrus and the amygdala which are located in the medial of the temporal lobe is named as Mesial Temporal Lobe Epilepsy (MTLE), while another one arising in the neocortex of the temporal lobe is named as Lateral Temporal Lobe Epilepsy (LTLE) (Alarcon *et al.*, 2001; Burgerman *et al.*, 1995).

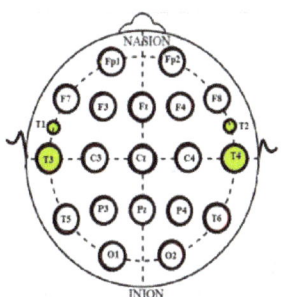

Figure 3.2: The 10-20 electrode placement system of 21 electrodes.

The anatomy of temporal lobe in left hemisphere brain, and the electrodes placed in anterior, middele and posterior temporal regions are illustrated in Fig. 3.3. In MTLE interictal epileptic discharges (IEDs) are picked up by T1, T2, and T3, T4 electrodes over the anterior and mesial temporal lobe while in LTLE maximum IEDs localized in T5 and T6 electrodes over the posterior (neocortical) temporal lobe (Pfander *et al.*, 2002; Hamer *et al.*, 1999). Moreover the TLE of neocortical onset is not well

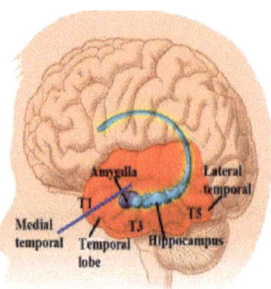

Figure 3.3: The anatomy of temporal lobe in left hemisphere brain and its corresponding electrodes placement in anterior (T1), mesial (T3) and posterior (T5) regions is highlighted.

understood or defined. Also, the LTLE seizures may quickly spread to ipsilateral temporal structures at the earliest and can be picked-up by electrodes over anterior and mesial temporal region (Brekelmans *et al.*, 1995; Fried, 1997; Ebersole and Pacia, 1996). As this research work considered only TLE, the EEG of four electrodes T1 and T3 from left, and T2 and T4 from right sides of the scalp as highlighted in Fig. 3.2 were selected to pickup the seizure onset from mesial and lateral temporal regions.

3.1.3 Sampling rate conversion

The EEG data are instantaneous values recorded by using 32 channels digital video-EEG (vEEG) system. The EEG data collected from two hospitals were recorded at two different sampling rates. In dataset_1 each channel was sampled at the rate of 400 samples per second. In dataset_2 each channel was sampled at the rate of 256 samples per second. To ensure the uniformity of sliding window length for both the datasets, two possible ways of sampling rate conversion were performed. First, dataset_2 with 256 samples per second was up-sampled to 400 samples per second by the factor of (25/16). The actual number of samples are interpolated by the factor of 25 and decimated by the factor of 16. Prior to decimation, the interpolated signal was filtered with 8th order Chebyshev Type I lowpass filter to mitigate the distortion caused by frequency aliasing. Alternately, dataset_1 with 400 samples per second was down-sampled to 256 samples per second by the factor of (16/25) to ensure the uniform window length for dataset_1 and dataset_2.

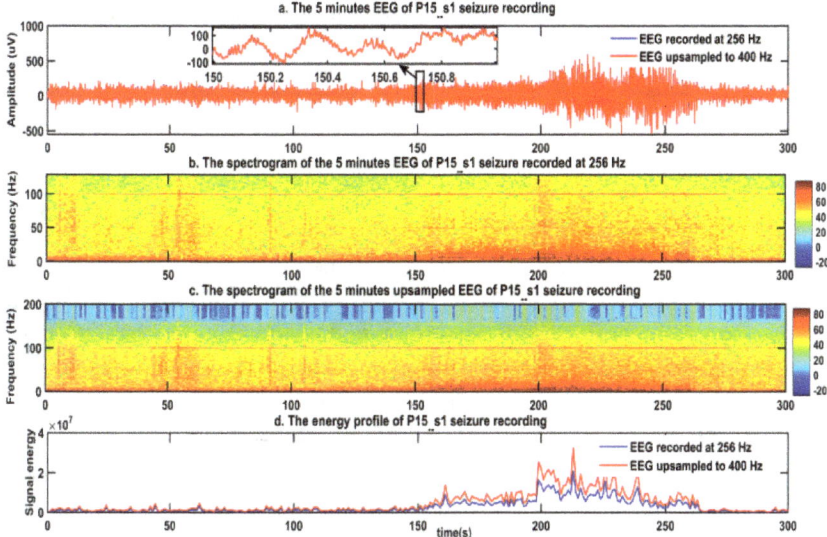

Figure 3.4: The amplitude profile of P15_s1 seizure recorded at 256 samples per second correlates with EEG upsampled to 400 samples per second. The spectral energy distribution of the recorded EEG shows cutoff frequency of 128 Hz and up-sampled signal shows the cutoff frequency of 200 Hz.

The seizure P15_s1 from dataset_2 was up-sampled to 400 Hz. The comparison of amplitude, spectral, and energy profile of actual and up-sampled 5 minutes EEG is given in Fig. 3.4. The amplitude profile of 5 minutes up-sampled EEG (red in color) finely coincide with the actual EEG (blue in color) is shown in Fig. 3.4(a). The spectrogram of actual and up-sampled 5 minutes EEG is shown in Fig. 3.4(b), and Fig. 3.4(c) respectively. The spectrogram of the recorded EEG of P15_s1 seizure shows the cutoff frequency of 128 Hz in Fig. 3.4(b). The spectrogram of up-sampled EEG of seizure P15_s1 shows the cutoff frequency of 200 Hz in Fig. 3.4(c). The energy distribution of recorded (blue in color) and up-sampled (red in color) EEG are shown in Fig. 3.4(d).

Ensuring the information refinement, the energy profile of up-sampled EEG follows the actual EEG with slightly more amplitude due to the addition of frequency components more than 128 Hz. Similarly, to verify the down-sampling conversion, P1_s1 seizure sample was down-sampled to 256 Hz. The EEG of P1_s1 seizure sampled at 400 Hz was passed through the lowpass filter with the cutoff frequency of 128 Hz to mitigate the distortion caused by frequency aliasing. The lowpass filtered signal

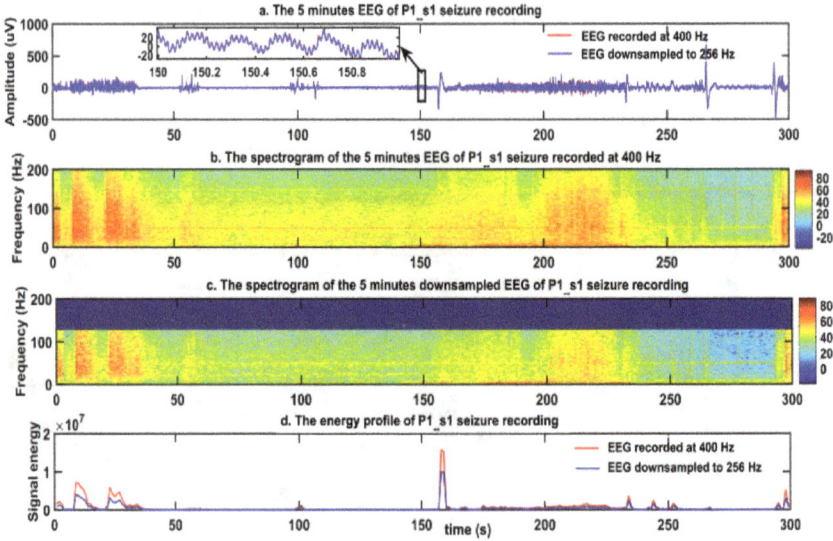

Figure 3.5: The amplitude profile of P1_s1 seizure recorded at 400 samples per second correlates with EEG down-sampled to 256 samples per second. The energy distribution of spectral components of recorded EEG shows cutoff frequency of 200 Hz whereas down-sampled signal with 128 Hz. The energy profile of down-sampled EEG (blue in color) and the recorded EEG (red in color) is shown.

was down-sampled to obtain 256 samples per second. The comparison of amplitude, spectral, and energy profile of actual and down-sampled EEG is given in Fig. 3.5(a) to Fig. 3.5(d). The amplitude profile of down-sampled EEG (blue in color) finely coincide with the raw EEG (red in color) is shown in Fig. 3.5(a). The spectrogram of actual and down-sampled EEG is shown in Fig. 3.5(b), and Fig. 3.5(c) respectively. The spectral energy distribution of recorded EEG shows the maximum frequency component of 200 Hz (half of the sampling frequency) in Fig.3.5(b). The down-sampled EEG of seizure P1_s1 shows the removal of frequency components above 128 Hz in Fig. 3.5(c). The distribution of energy of recorded and down-sampled EEG are shown in Fig. 3.5(d). The energy profile of recorded EEG (red in color) is correlated with the down-sampled EEG (blue in color) with reduced amplitude which reveals the loss of information above 128 Hz.

Among the two sampling rate conversions, up-sampling the 256 Hz signal retains the original EEG whereas down-sampling the 400 Hz signal discards the information beyond 128 Hz. Moreover, the original signal can be reconstructed from up-sampled signal where as the signal lost in down-sampling, can not be retrieved. In this work, the dataset_2 was up-sampled to 400 Hz to achieve the uniformity across seizures and sliding window length for both the datasets.

3.1.4 EEG Filtering

The scalp EEG signal is composed of low-amplitude voltage oscillations generated by cortical neurons in the range of microvolts. The recorded EEG contaminated by different noise sources, and mixed with biological signals such as ECG, EOG and respiratory signals. The most common source of artifacts are direct current (DC) potential generated by charges accumulation (half-cell potential) on the metal surface which induce the voltage in the range of millivolts, and 50 Hz supply line interferences. If these power source artifacts are not removed, these will be amplified largely and contaminate the EEG signal. Hence before signal analysis, DC potential and 50 Hz supply line interferences are removed using high-pass filters (HPF), and notch filters respectively. The analog EEG filters are implemented using Resistor-Capacitor (R-C) circuits with proper selection of time-constant ($\tau = RC$). The time-constant value of the filter determines the choice of resister and capacitor values. The high-pass filters are designed with the cut-off frequencies ($F_{cut-off} = 1/(2\pi\tau)$) of 0.1, 0.3, 1, 3 or 10 Hz corresponding to the time-constants of 1.6, 0.53, 0.16, 0.05 or, 0.016 seconds. Also, these are the standard values for digital high-pass filters which are most suitable for scalp recorded EEG frequencies.

Among these, the most common setting for high-pass filter is 1 Hz which preserves the most low-frequency information by largely eliminating the drift of baseline due to DC potential along with biological signals resulting from respiration. Moreover, increasing the cut-off frequency of high-pass filter flatten the slow oscillations while leaving the high frequency components intact. In this work, combination of 1 to 200 Hz frequency components are used to represent the EEG to extract the features. The EEG was passed through a second-order high-pass filter to obtain 1 to 200 Hz band limited signals. The second-order notch filters were used to remove the presence of 50 Hz, and its harmonics 100 Hz and 150 Hz supply-line interferences. For this study, all the 35 seizure recordings were filtered using same filter settings to attain patient independent

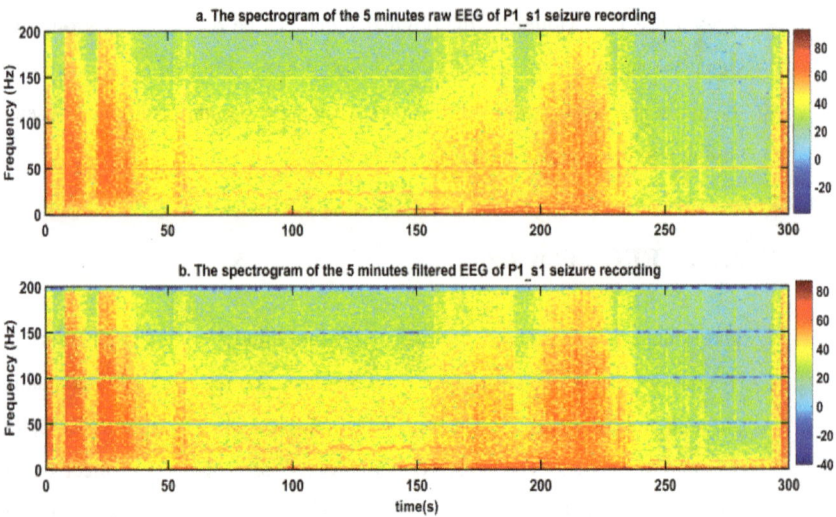

Figure 3.6: The spectrogram of raw and filtered EEG of seizure P1_s1 shows the frequency activities of 5 minutes EEG. The 50 Hz supply line frequency and, its harmonics of 100 Hz, 150 Hz were removed.

ability. The spectrogram of raw and filtered 5 minutes EEG of P1_s1 seizure recording is shown in Fig. 3.6. The spectrogram of unfiltered EEG in Fig. 3.6(a) shows the presence of 50 Hz, 100 Hz and 150 Hz supply-line interferences. The spectrogram of filtered EEG in Fig. 3.6(b) shows the removal of 50 Hz and its harmonic frequencies of 100 Hz,and 150 Hz.

3.1.5 Electrical vs Clinical onset of seizures

For 29 scalp video-EEG recordings of 18 patients (P1 to P21), the electrical onset and clinical onset of seizures were visually identified by three Neurologists. Based on the electrical and clinical onset, the seizure recordings were divided into five groups. In the first group of nine seizures, the electrical onset preceded the clinical onset with mean latency of 54 seconds (ranging from 42 seconds to 69 seconds with the median of 55 seconds). Fig. 3.7 shows an example from the first group of the onset of seizure P1_s1 in the right temporal lobe, for four channels of the scalp EEG. The electrical manifestation of seizure started at 138 seconds in T2, the right temporal region with 4 Hz theta band oscillations, and spread into ipsi-lateral hemisphere is picked-up by T4 electrode. The amplitude of EEG is increasing gradually and producing 6 Hz sustained oscillations in upper theta band. Later, the seizure slowly propagate to contra-lateral hemisphere is picked up by T1 and T3 electrodes. The large potential at 157 seconds is picked-up by all the four electrodes and the changes are more dominant in T2 electrode.

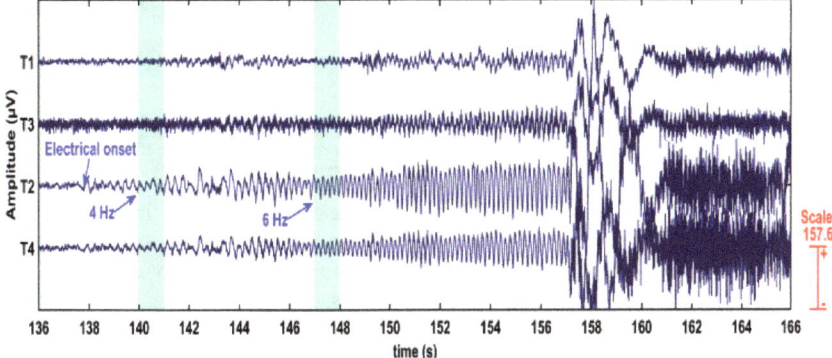

Figure 3.7: Scalp-recorded EEG of the selected electrodes in P1_s1 seizure recording. The seizure onset was visually identified in the temporal region, and 4 Hz rhythmic theta activity commenced at the T2 electrode (arrow) at 138 seconds.

In the second group of four seizures, the electrical onset preceded the clinical onset with a mean latency of 37 seconds (ranging from 35 seconds to 39 seconds with the median of 38 seconds). A typical example from the second group of the onset of seizure P15_s1 in the right temporal lobe, is shown in Fig. 3.8. The seizure started at 145 seconds in T2, the right temporal region with 4 Hz theta band oscillations, and slowly spread into ipsi-lateral hemisphere is picked-up by T4 electrode. The amplitude of EEG is increasing gradually and producing 7 Hz sustained oscillations in upper theta band. Later, the seizure propagates to contra-lateral hemisphere, and the clinical seizure begins at 180 seconds can be noticed. The spectrogram of 5 minutes filtered EEG of P15_s1 seizure picked-up by T2 electrode is shown in Fig. 3.9. The theta band spectral activity is clearly visible at the electrical onset of seizure marked at 145 seconds (arrow).

In the third group of four seizures, the electrical onset preceded the clinical onset with the mean latency of 23 seconds (ranging from 21 seconds to 28 seconds with the median of 22 seconds). The example for the third group of the onset of seizure P7_s2 in the left temporal lobe, is shown in Fig. 3.10. The seizure started at 157 seconds in T1, the left temporal region with clear 7 Hz theta band oscillations, and propagate

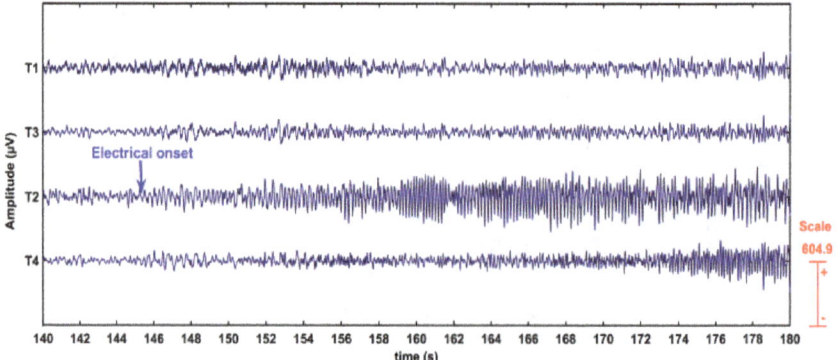

Figure 3.8: Scalp-recorded EEG of the selected electrodes in the P15_s1 seizure recording. The electrical onset of seizure was visually identified in the temporal region, and 7 Hz rhythmic theta activity commenced at the T2 electrode (arrow) at 145 seconds.

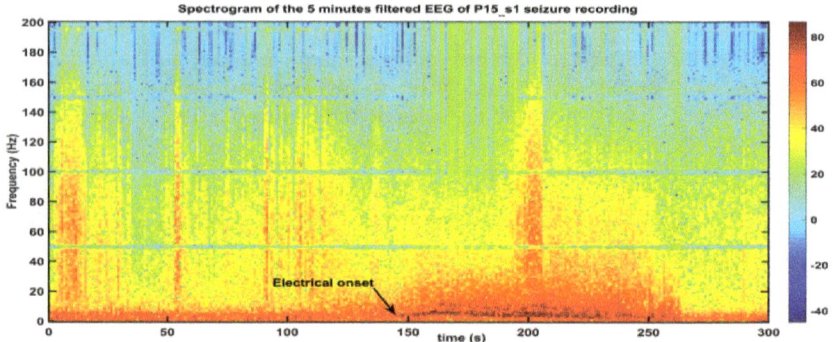

Figure 3.9: The spectrogram of the 5 minutes filtered EEG of P15_s1 seizure recorded at T2 electrode.

to ipsi-lateral hemisphere is picked-up by T3 electrode. The amplitude of EEG is increasing gradually and producing 7 Hz sustained oscillations in upper theta band. Later the seizure propagates to contra-lateral hemisphere, spike wave discharges are picked-up by T2 and T4 electrodes at 171 seconds, and the clinical seizure begins at 180 seconds. The spectrogram of 5 minutes filtered EEG of P7_s2 seizure picked-up by T1 electrode is shown in Fig. 3.11. The theta band spectral activity is clearly visible at the electrical onset of seizure marked at 157 seconds (arrow).

In the fourth group of seven seizures, the electrical onset preceded the clinical onset with a mean latency of 16 seconds (ranging from 13 seconds to 19 seconds with the median of 17 seconds). An example from the fourth group of the onset of

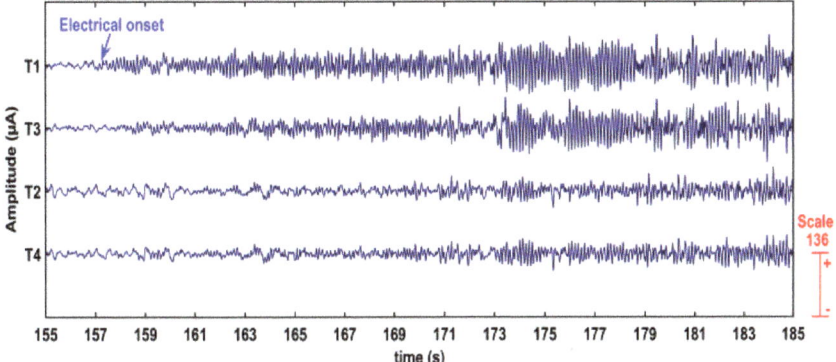

Figure 3.10: Scalp-recorded EEG of the selected electrodes in the P7_s2 seizure. The electrical onset of P7_s2 was visually identified in the left temporal region, and clear 7 Hz rhythmic theta activity commenced at the T1 electrode (arrow) at 157 seconds.

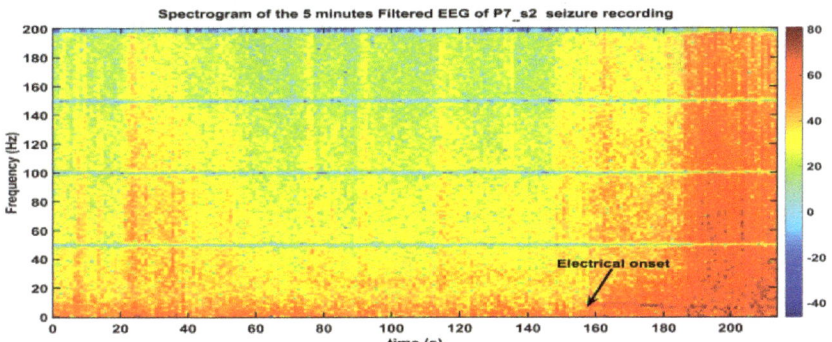

Figure 3.11: The spectrogram of the 5 minutes filtered EEG of P7_s2 seizure recorded at T1 electrode.

seizure P17_s1 in the right temporal lobe, is shown in Fig. 3.12. The seizure started at 163 seconds in T2 electrode. The seizure begins with spike discharges with increasing amplitude followed by 7 Hz sustained neuronal oscillations in the theta rhythms. The seizure activity picked-up by T4 electrode which is little away from epileptic-foci. The spectrogram of 5 minutes filtered EEG of P17_s1 seizure picked-up by T2 electrode is shown in Fig. 3.13. The spectral activity of the seizure starting from its electrical onset marked at 163 seconds (arrow) is clearly visible.

In the fifth group of five seizures, the electrical onset preceded the clinical onset with a mean latency of 5 seconds (ranging from 3 seconds to 6 seconds with the median of 5 seconds). An example from the fifth group of onset of seizure P2_s1 in

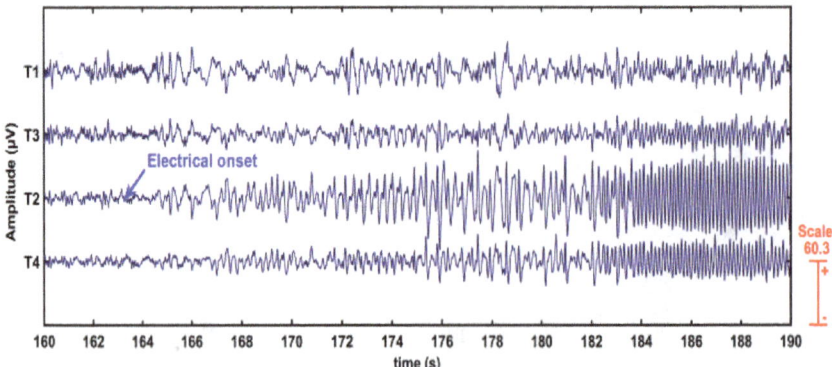

Figure 3.12: Scalp-recorded EEG of the selected electrodes in the P17_s1 seizure. The electrical onset of P17_s1 was visually identified in the right temporal region at T2 electrode (arrow) at 163 seconds.

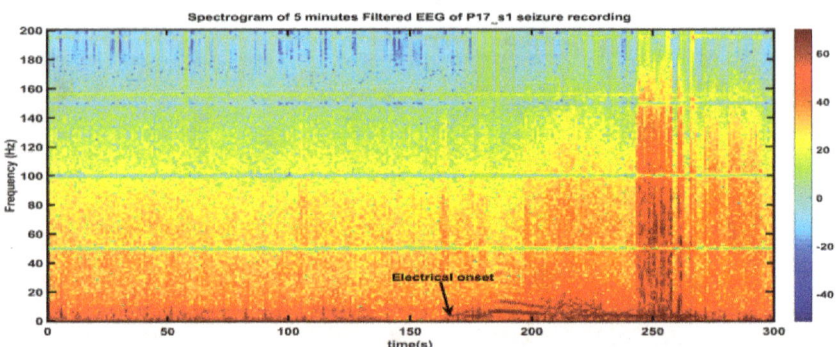

Figure 3.13: The spectrogram of the 5 minutes filtered EEG of P17_s1 seizure recorded at T2 electrode.

the right temporal lobe, started at 174 seconds at T2 electrode is shown in Fig. 3.14. The seizure begins with 5 Hz neuronal oscillations followed by spike discharges with increasing amplitude is dominant in T2 electrode. The seizure activity is also picked-up by T4 electrode which is little away from epileptic-foci. Later the seizure propagates to contra-lateral hemisphere and developed as generalized seizure. The summary of the onset latency of five groups of seizures is tabulated in Table 3.2.

After understanding the EEG patterns of electrical and clinical onset of the above mentioned five clusters of seizure recordings, an automated detection algorithm was developed to identify the electrical onset of seizure. In the process of developing automated detection algorithm, each seizure recording was divided into four seconds segments with each segment overlapping the previous segment. The detailed

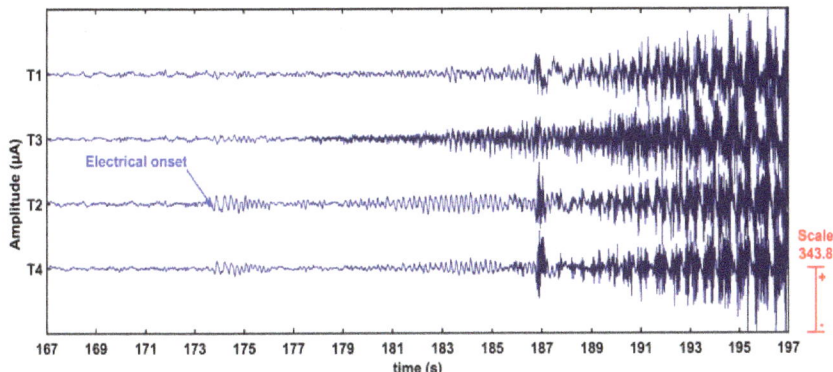

Figure 3.14: Scalp-recorded EEG of the selected electrodes in the P2_s1 seizure. The electrical onset of seizure was visually identified in the right temporal region at T2 electrode (arrow) at 174 seconds.

Table 3.2: The summary of the onset latency of five groups of seizures

Seizures Group	No. of Seizures	Latency (s)		
		Range	Mean	Median
Group 1	9	42 to 69	54	55
Group 2	4	35 to 39	37	38
Group 3	4	21 to 28	23	22
Group 4	7	13 to 19	16	17
Group 5	5	3 to 6	5	5

Note: Latency - The difference in time, the electrical onset preceded the clinical onset

preprocessing pipeline is depicted by flowchart in Fig. 3.15. For each segment a set of features were extracted to characterize that segment EEG data. In this way for each seizure recording, a set of features were extracted at an interval of one second. These features set extracted for every one second were fed into the classifiers for automated identification of electrical onset of seizure. The electrical and clinical onset of six seizure recordings of four patients (P22, P23, P24, P25) were also identified by the Neurologists. The features extracted from these six seizure recordings which are not involved in the development of automated detection algorithm, were used for testing the proposed system. Finally, the automated identification of electrical onset of all the

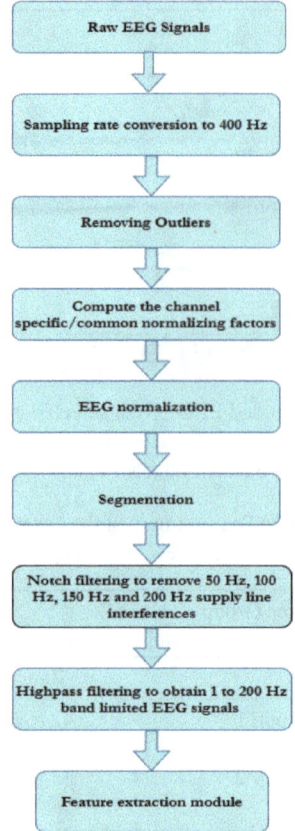

Figure 3.15: The detailed pre-processing pipeline is depicted by flowchart.

seizures by the classifiers were compared with the electrical onset visually identified by the Neurologists. The detection performance of the classifiers was investigated, and the suitability of the proposed system for clinical routine was analyzed.

3.1.6 EEG Normalization

Due to the interconnection between billions of neurons, the recorded EEG is more complex and random in nature. Moreover, EEG dynamics varies across patients, and seizures of the same patient. Because of the amplitude of each patient EEG varied, the signals were normalized to make the amplitudes relatively comparable across patients. The EEG data of four selected channels were normalized using two normalization schemes: Scheme 1 - Each recording was normalized, using the channel specific maximum and minimum values computed from that seizure recording; Scheme 2 - Each recording was normalized, using the channel specific average maximum, and average minimum values computed from all 29 seizure recordings.

Normalization Scheme 1

The maximum and minimum value of each channel in the 29 seizure recordings were computed and tabulated in Table 3.3. The EEG samples of each recording was normalized from 0 to 1 range using the channel specific maximum and minimum values as specified in Eq. (3.1),

$$x_{norm} = \frac{x_i - x_{minimum}}{x_{maximum} - x_{minimum}}, \quad 0 \leq x_{norm} \leq 1 \qquad (3.1)$$

where x_{norm} is the normalized value of x, x_i is the instantaneous value of x, and $x_{minimum}$ and $x_{maximum}$ are the channel specific minimum and maximum values computed from that seizure recording.

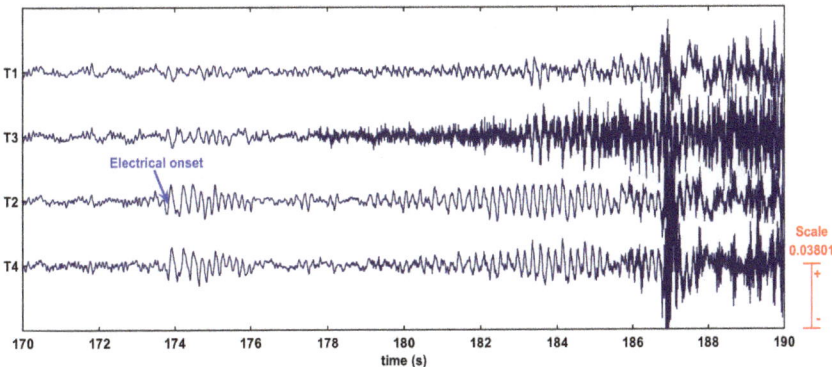

Figure 3.16: The EEG of seizure P2_s1 normalized using channel specific maximum and minimum value computed from the seizure recording (Scheme 1). The seizure electrical onset at 174 seconds in T2 from right temporal region.

The scalp recorded EEG of seizure P2_s1 normalized using channel specific maximum and minimum value computed from that seizure recording is shown in Fig. 3.16. The seizure electrical onset at 174 seconds in T2 electrode from right temporal region can be observed.

Table 3.3: Maximum and minimum value of four channels in 29 seizure recordings and their average values

Seizure ID	Maximum value (μV)				Minimum value (μV)			
	T1	T3	T2	T4	T1	T3	T2	T4
P1_s1	526	398	847	369	-599	-636	-636	-477
P1_s2	584	605	576	709	-1145	-617	-358	-772
P1_s3	400	221	212	346	-385	-341	-274	-478
P2_s1	1723	1235	1587	1118	-1741	-1580	-1775	-1341
P2_s2	1140	1010	1704	613	-1350	-1702	-1732	-895
P2_s4	1192	836	1510	582	-1512	-1561	-1614	-1342
P3_s1	1394	1472	1065	1712	-1498	-1715	-1341	-1577
P4_s1	103	114	126	100	-112	-222	-115	-153
P5_s1	454	347	427	336	-320	-277	-621	-330
P7_s1	1394	1472	1065	1712	-1498	-1715	-1341	-1577
P7_s2	895	909	669	952	-1079	-1025	-806	-1021
P7_s3	516	524	761	1649	-876	-1154	-906	-920
P8_s1	442	400	670	544	-388	-525	-619	-627
P8_s3	2023	498	528	528	-1198	-926	-786	-930
P11_s1	1394	1472	1581	1712	-1498	-1715	-1341	-1577
P11_s2	1394	1472	1581	1712	-1498	-1715	-1341	-1577
P11_s3	1394	1472	1065	1712	-1498	-1715	-1341	-1577
P12_s1	464	1835	227	258	-435	-1662	-182	-242
P13_s1	350	367	350	315	-536	-527	-392	-444
P14_s1	964	193	1721	164	-575	-202	-1127	-238
P15_s1	670	546	468	479	-543	-569	-567	-616
P16_s1	293	182	353	203	-235	-257	-401	-253
P17_s1	336	206	335	138	-637	-301	-389	-353
P17_s2	341	197	421	151	-468	-418	-564	-366
P18_s1	748	1619	430	1568	-254	-400	-404	-168
P19_s1	810	824	193	182	-1379	-869	-369	-425
P20_s1	169	173	331	284	-253	-181	-343	-328
P20_s2	107	127	211	173	-192	-102	-322	-187
P21_s1	327	344	301	270	-425	-646	-287	-390
Average	**777**	**726**	**735**	**710**	**-832**	**-871**	**-769**	**-730**

Normalization Scheme 2

In this scheme, the maximum and minimum values computed from 29 seizure recordings are varying largely across the patients. Moreover, using seizure specific normalization values, the selection of maximum and minimum value for the upcoming seizure is unpredictable. Hence, to make the EEG comparable across the patients, and to detect the electrical onset of upcoming seizure, one common normalization factor was obtained. The maximum values and minimum values of T1, T3, T2, and T4 channels in all the 29 seizure recordings were computed. The average of the 29 maximum values and average of the 29 minimum values of each channel were used to normalize the

5 minutes EEG samples in all 29 seizure recordings, as specified in Eq. (3.2),

$$x_{channel\_norm} = \frac{x_{channel\_i} - Avg.channel\_minimum}{Avg.channel\_maximum - Avg.channel\_minimum}, \quad 0 \leq x_{norm} \leq 1 \quad (3.2)$$

where $x_{channel\_norm}$ is the normalized value of x from specific channel, x_i is the instantaneous value of x from specific channel, and $Avg.channel\_minimum$ and $Avg.channel\_maximum$ are the channel specific average minimum and average maximum values computed from all 29 seizure recordings.

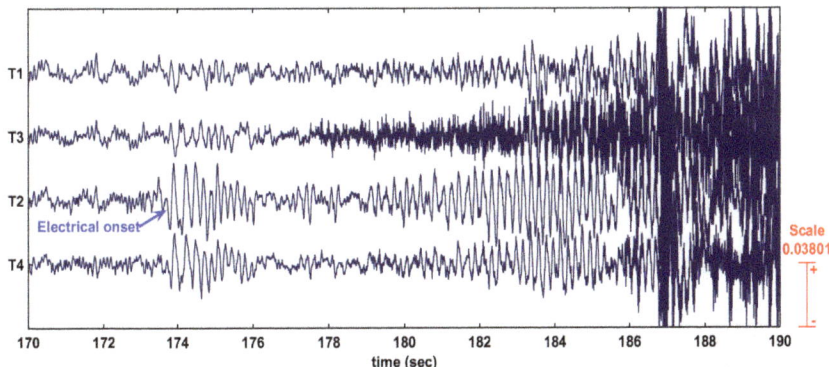

Figure 3.17: The EEG of seizure P2_s1 normalized using average maximum and minimum value computed from 29 seizure recordings (Scheme 2). The seizure electrical onset at 174 seconds from right temporal T2 electrode.

The scalp recorded EEG of seizure P2_s1 normalized using average maximum and average minimum value computed from 29 seizure recordings is shown in Fig. 3.17. From Table 3.3, the maximum value of 5 minutes recording of seizure P2_s1 is larger than the averaged maximum value of P2_s1, and minimum value is smaller than the averaged minimum value. Comparing the EEG of P2_s1 normalized by two schemes as shown in Fig. 3.16 and Fig. 3.17, EEG normalized using Scheme 2 follows the same patterns as Scheme 1 with increased amplitude profile. Similarly, the EEG of seizure P4_s1 normalized by using Schemes 1 and 2 is presented in Fig. 3.18 and Fig. 3.19 respectively. The seizure specific maximum and minimum value of P4_s1 recording is smaller than the average maximum and average minimum values. Hence the amplitude of EEG normalized using Scheme 1 appears larger than the amplitude of

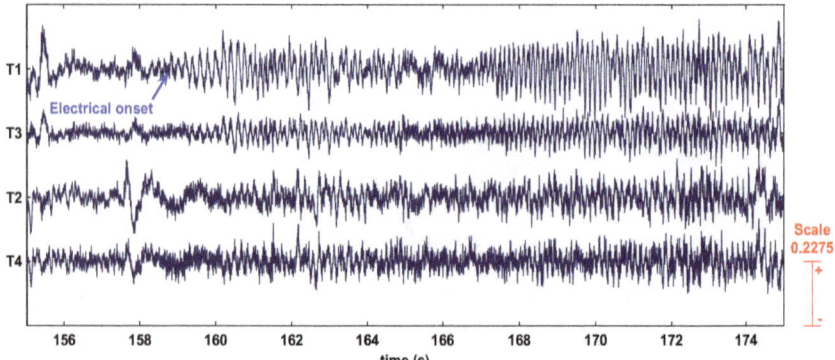

Figure 3.18: The EEG of seizure P4_s1 normalized using channel specific maximum and minimum value computed from the seizure recording (Scheme 1). The seizure electrical onset at 159 seconds from left temporal lobe T1 electrode.

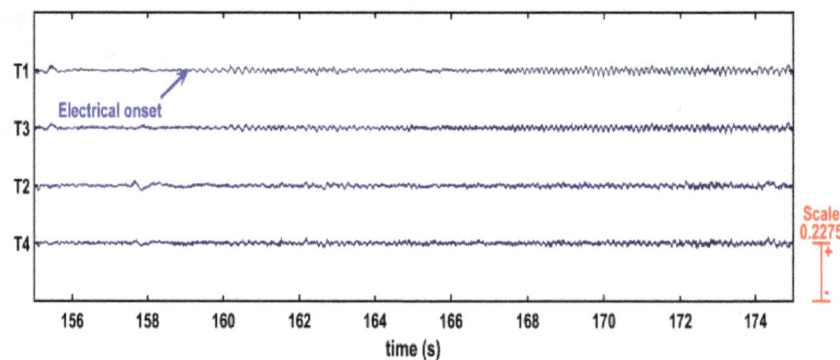

Figure 3.19: The EEG of seizure P4_s1 normalized using average maximum and average minimum values computed from 29 seizure recordings (Scheme 2). The seizure electrical onset at 159 seconds from left temporal lobe T1 electrode.

EEG normalized using Scheme 2. However the normalization schemes did not show much effect in identifying the electrical onset of seizures by the classifiers, which will be discussed further in the following chapters.

3.2 Summary

Seizures are the hyper synchronous activity of neuronal network. The clinical symptoms accompanying the seizure varies based on its cerebral site of onset as well as its patterns spread to surrounding brain areas. In this study, the scalp recorded electroencephalogram which measure the aggregate electrical activity of populations of neurons were used to detect the onset of temporal lobe seizures. The EEG of 35 temporal lobe epileptic seizures were filtered to obtain 1 to 200 Hz band limited signals. The amplitude of the EEG varies significantly across individuals. Hence, the seizure recordings were normalized using two different normalization schemes to make the EEG comparable across individuals. In the following chapters, the normalized EEG is used to extract the features to detect the electrical onset of seizures. The time and frequency domain features extraction, and their ability to distinguish the normal and seizure features samples are described in the subsequent Chapter.

CHAPTER 4

DESIGN OF SEIZURE DETECTION SYSTEM USING NORMALIZATION SCHEME 1

The automated electrical seizure onset detection system is designed using the features extracted from EEG normalized using the maximum and minimum values of the individual seizure recording (Normalization Scheme 1). A set of time and frequency domain features such as signal energy, approximate entropy, sample entropy, mean, variance, skewness, kurtosis, spectral energy, wavelet energy, and spectral entropy are extracted for all 29 seizure recordings. A sliding window technique is adapted to extract the features of 5 minutes EEG of each seizure recording. The 5 minutes EEG is segmented into M (M=2,3,4,5) seconds windows with M-1 seconds overlapping with previous window. The characteristics of each 5 minutes EEG is analyzed using M equal to 2 seconds, 3 seconds, 4 seconds and 5 seconds window, and optimal length for sliding window is chosen. Then the optimal window sliding over 5 minutes EEG is used to extract the features to design the automated electrical seizure onset detection system. In this chapter the suitability of selected features that helpful to detect the electrical onset of seizures are examined. The performance of five classifiers – Linear Discriminant Analysis (LDA), Naive Bayes (NB), Decision Tree (DT), Support Vector Machine (SVM), and K Nearest Neighbors (KNN) – trained using significant features to distinguish the difference between the normal, and seizure features samples are investigated.

4.1 Feature extraction

A set of 29 seizures recorded from 18 patients were used for the design of automated seizure detection system. From each seizure recording, only four channels T1, T3 and T2, T4 from left and right hemisphere of the brain were selected. As explained in Section 3.1.6, the EEG samples of each seizure recording were normalized using the channel specific maximum and minimum values computed from that seizure recording.

The Scheme 1 normalized EEG was passed through the second order filter to obtain the 1 to 200 Hz band limited signal. The supply line interferences were removed using notch filters with 50 Hz, 100 Hz, 150 Hz and 200 Hz center frequencies. For each 5 minutes 1 to 200 Hz band limited EEG, a set of time and frequency domain features were calculated. The features calculation procedure and formule are stated in the following sections.

4.1.1 Signal Energy

In time domain, the EEG signals were analyzed to track the changes in amplitude of brain potentials. The signal energy was measured as the sum of squared instantaneous voltage of EEG samples. The energy of the signal was calculated by Eq. (4.1),

$$E_k = \sum_{i=(k-1)D+1}^{(M+k-1)D} |x_i|^2 \qquad (4.1)$$

where E_k is the signal energy in the kth segment, k is the segment index, i is the sample index in the segment, M is the length of sliding window in seconds varied from 2 to 5, D is the sampling rate, and x_i is the instantaneous sample value of the segment. In this thesis work, the signal energy was calculated for the entire spectrum of 1 to 200 Hz.

4.1.2 Entropy

The entropy is a measure of irregularity, and complexity of the stochastic signals. In this thesis work, two entropy measures – approximate entropy (ApEn) and sample entropy (SampEn) were selected to quantify the amount of irregularity within a finite length EEG time series $x(n) = [s(1)\ s(2)\ ...\ s(N)]$. It consists of N data samples $s(1)$, $s(2)$, ... $s(N)$ equally spaced in time. The time series with more regular patterns produce low entropy values and vice versa. The calculation of these entropy values of the given time series mainly depending on three parameters namely the size of matching pattern (p), threshold level for matching (r) and the length of time series (N). In the previous studies, these entropy measures were examined using different values of these parameters (p = 1, 2, 3; r = 0.1, 0.15, 0.2, 0.3; N = 173, 256, 512, 1024, 2048 samples) for the automated classification of normal, interictal, and ictal EEG samples (Kannathal et al., 2005; Srinivasan et al., 2007; Song et al., 2012). In this research work the entropy values were calculated for each M (2 to 5) seconds window with the set of values for p=2,3 and r=0.1, 0.2, 0.3.

Approximate entropy

Pincus (1991) developed the ApEn to estimate the predictability within the time series $x(n)$. For each window, the ApEn was calculated for the set of p=2 and r=0.1 values using the following steps:

1. For each window, N samples are divided into the sequence of N-p+1 p-dimensional vectors $[y(1), y(2), y(3), ... y(N-p+1)]$. Each vector is defined as $y(i) = [s(i)\ s(i+1)\ ...\ s(i+p-1)]$ for 'i' varying from 1 to N-p+1.

2. For each window, the first p-dimensional vector $y(1) = [s(1)\ s(2)]$ is compared with all p-dimensional vectors y(j), $\forall\ j\ 1 \leq j \geq (N-p+1)$. Following that, the second p-dimensional vector $y(2) = [s(2)\ s(3)]$ is compared with all p-dimensional vectors y(j). Similarly, all the p-dimensional vectors y(i) are compared with all p-dimensional vectors y(j). For each p-dimensional vector, the distance d[y(i), y(j)] between the two vectors y(i) and y(j) is calculated as given by Eq. 4.2. If $d[y(i), y(j)]$ is within the range of threshold value (r=0.1), then the vector y(j) is similar to vector y(i).

$$d[y(i), y(j)] = max\ (|y(i+k-1) - y(j+k-1)|),\ \ k = 1, 2, \cdots, p \tag{4.2}$$

3. For each window the number of similar vectors for each vector y(i) is counted and the conditional probability $C_i^p(r)$ representing the similarity between y(i) and y(j) vectors is given by Eq. 4.3,

$$C_i^p(r) = \frac{number\ of\ y(j)\ such\ that\ d[y(i),\ y(j)] \leq 0.1}{(N-p+1)} \tag{4.3}$$

4. For each y(i) the alike vectors y(j) which are fell within the threshold level are calculated using $C_i^p(r)\ \forall\ i,\ 1 \leq i \geq (N-p+1)$. For the selected 'p' and 'r' values, the total sum of the logarithm of conditional probabilities of all N-p+1 vectors divided by N-p+1 is computed as given in Eq. 4.4.

$$\Phi^p(r) = (N-p+1)^{-1} \sum_{i=1}^{N-p+1} \ln C_i^p(r) \tag{4.4}$$

5. Similarly, keeping r the same value, p is incremented by 1 and repeated the steps from 1 to 4 to obtain $\Phi^{p+1}(r)$ as given in Eq. 4.5.

$$\Phi^{p+1}(r) = (N-p)^{-1} \sum_{i=1}^{N-p} \ln C_i^{p+1}(r) \qquad (4.5)$$

6. The Approximate entropy (ApEn) is calculated by subtracting the conditional probabilities of p+1 from p embedding data sequences as given in Eq. (4.6),

$$ApEn = \Phi^p(r) - \Phi^{p+1}(r) \qquad (4.6)$$

Sample entropy

The ApEn inherently introduce bias towards the regularity, by counting the self-match of the vectors. Also, selection of parameters may flip the values, and fail to produce consistent result (Yentes *et al.*, 2012). Richman and Moorman (2000) introduced SampEn to counteract the shortcomings of ApEn calculation. SampEn does not count a self-match, thus eliminating the bias towards regularity.

Similar to ApEn, the SampEn calculates the conditional probability for p embedding dimension data sequences $C_i^p(r) \; \forall \; i, 1 \leq i \geq (N-p+1)$ by comparing $y(i)$ with $y(j)$, $\forall \; y(i) \neq y(j)$. Similarly the SampEn calculates the conditional probability for p+1 embedding dimension $C_i^{p+1}(r) \; \forall \; i, \; 1 \leq i \geq (N-p)$ by comparing $y(i)$ with $y(j)$ $\forall \; y(i) \neq y(j)$. Finally the SampEn is calculated as the natural logarithm of the sum of conditional probabilities as given in Eq. 4.9,

$$SampEn = -\ln \frac{A}{B} \qquad (4.7)$$

where,

$$A = \frac{C_i^{p+1}(r)}{N-p}$$
$$B = \frac{C_i^p(r)}{N-p+1} \qquad (4.8)$$

$$(4.9)$$

In this research work, for the detection of electrical onset of seizure, these two entropy measures were calculated for the embedding dimension of p = 2,3 and threshold values of r = 0.1,0.2,0.3 × (SD) of time series. The standard deviation (SD) of the data within each windowed EEG signal was considered for defining the identical patterns for p and p+1 embedding dimension vectors.

4.1.3 Statistical measures

The statistical analysis uses 1st, 2nd, 3rd and 4th order moments to analyse the distribution of EEG data samples. The first order moment which represents the average amplitude of the signal was calculated using Eq. (4.10),

$$\hat{x}_k = \frac{1}{N} \sum_{i=(k-1)D+1}^{(M+k-1)D} x_i \qquad (4.10)$$

The variance is the second order moment to test the spread of samples about the mean. The variance of time-series was calculated using Eq. (4.11),

$$\sigma_k^2 = \frac{1}{N} \sum_{i=(k-1)D+1}^{(M+k-1)D} (x_i - \hat{x}_k)^2 \qquad (4.11)$$

The skewness and kurtosis are 3rd and 4th order moments to get the information about the shape of the distribution. The skewness was calculated using Eq. (4.12) to test the length of tail on both sides of the distribution.

$$s_k^2 = \frac{1}{N} \sum_{i=(k-1)D+1}^{(M+k-1)D} \frac{(x_i - \hat{x}_k)^3}{\sigma_k^3} \qquad (4.12)$$

If the value of skewness is negative, the distribution is skewed left; otherwise right skewed. The kurtosis was used to test the peakyness of the distribution. The kurtosis was calculated using Eq. (4.13),

$$k_k^2 = \frac{1}{N} \sum_{i=(k-1)D+1}^{(M+k-1)D} \frac{(x_i - \hat{x}_k)^4}{\sigma_k^4} \qquad (4.13)$$

where k is the segment index, M is the length of sliding window in seconds varied from 2 to 5, D is the sampling rate, N is the number of samples in each window, i is the sample index, x_i - instantaneous value of EEG, $\hat{x}_k$ - kth segment mean, σ_k^2- kth segment variance, s_k- kth segment skewness and k_k- kth segment kurtosis.

4.1.4 Spectral Energy

The brain produces different rhythmic patterns depending on mental state and task performance. The power of the EEG signal is mostly concentrated in low frequencies. During the evolution of a seizure, changes in the rhythmic activity occur. The seizure evolves with different morphologies and sustained oscillations in different frequency bands including delta, theta, alpha, beta, and showing amplitude depression, and poly-spike patterns along with large amplitude waves (Meier *et al.*, 2008). The frequency signatures during the transition from normal to ictal phases must be captured using frequency transformation tools to detect the seizure precursors. To identify the variations in these frequency rhythms, the spectral energy of the windowed EEG signal was calculated using fast Fourier transform (FFT) under the assumption that the EEG signal is statistically stationary within the finite window length. The spectral energy of the windowed EEG signal was calculated using Eq. (4.14),

$$E_s = \sum_{f_1}^{f_2} |X(\omega_i)|^2 \qquad (4.14)$$

where E_s is the spectral energy in the f_1– f_2 band of frequencies, and $X(\omega_i)$ is the Fourier coefficient of the signal at angular frequency ω_i. In this research work, the spectral energy was calculated in 1 to 3 Hz, 3 to 6 Hz, 6-12 Hz, 12-25 Hz, 25-50 Hz, 50 to 100 Hz and 100 to 200 Hz frequency bands to analyze the changes in brain rhythmic patterns.

4.1.5 Wavelet energy

During the evaluation and propagation, the seizure produces rhythmic spiking discharges in different frequency bands. The discrete wavelet transform was used to analyse the distribution of wavelet energy of the EEG signal. The mother wavelet Daubechies-4 (Indiradevi *et al.*, 2008; Osorio *et al.*, 1998; Gigola *et al.*, 2004) which mimic the shape of epileptiform discharges was used to analyse the EEG signal frequency components. The signal was decomposed into six levels to obtain the energy profile in the same frequency bands as identified in the Section 4.1.4, 1 to 3 Hz, 3 to 6 Hz, 6 to 12 Hz, 12 to 25 Hz, 25 to 50 Hz, 50 to 100 Hz and 100 to 200 Hz.

The jth level wavelet energy E_j, was calculated using Eq. (4.15),

$$E_j = \sum_{k=1}^{l(j)} |D_j(k)|^2 \qquad (4.15)$$

Where j is the coefficient level, k is the number of samples in the corresponding level, $l(j)$ is the length of jth level, D_j is the coefficient at level j.

4.1.6 Spectral entropy

During the normal state, the EEG signal is random in nature. But during the approach to the seizure the brain produces deterministic oscillations with high neural synchrony. The spectral entropy is a suitable measure for capturing the synchronous activity of the brain in frequency space. The major advantage of using spectral entropy is that the frequency band contributing the entropy is known and user-defined. The previous use of spectral entropy for detection (Kannathal et al., 2005) and prediction (Blanco et al., 2013) motivated to select this as one of the measures to detect the electrical onset of seizure. In the work of Blanco et al., the invasively measured spectral entropy showed variations in the high frequency bands from 32 to 128 Hz. Since the scalp EEG is used in this study, the oscillations within lower frequencies band were considered. The spectral entropy was calculated using Eq. (4.16),

$$S = \sum_{f_1}^{f_2} p(f_i) . log\left(\frac{1}{p(f_i)}\right) \quad \in 0, 1 \qquad (4.16)$$

where S is the spectral entropy in the frequency range f_1–f_2, and $p(f_i)$ is the power spectral density of f_i. Among the different types of entropy mentioned in the previous section, the spectral entropy provides flexibility in frequency band selection for entropy estimation. Utilizing the flexibility in frequency band selection, the spectral entropy were estimated for different frequency bands selected within 1 to 200 Hz. From the analysis, 3 to 12 Hz frequency band which correlates with temporal lobe ictal rhythms (Ebersole and Pacia, 1996) was chosen to calculate the spectral entropy feature.

4.2 Selection of optimal window length

The sliding window technique was adopted to extract the features of each 5 minutes seizure recording. The selection of window length renders the profile of characterizing features more meaningful for the seizure detection system design. During the progression of seizure, the EEG pattern changes to the maximum of 8 phases, each phase lasts for 2 to 3 seconds (Wu and Gotman, 1998; Dericioglu and Saygi, 2008; Fisher *et al.*, 2014). The optimal window length must be selected to capture these seizure patterns evolving over time. In this research work, windows of four different lengths, 2 seconds, 3 seconds, 4 seconds, and 5 seconds were used to obtain the characteristic changes of the EEG.

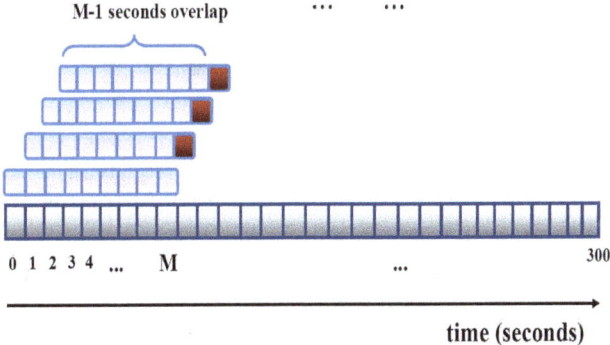

Figure 4.1: Sliding window technique adopted to extract the features of 5 minutes EEG.

The sliding window technique adopted to extract the features over 5 minutes of EEG is illustrated in Fig. 4.1. The M seconds predefined window is slid by 1 second, keeping M-1 seconds of data overlapping with the previous window to obtain the profile of the features. For all 29 seizure recordings, the features were extracted from the four selected T1, T2, T3, and T4 channels. Among the four selected window length, 2 seconds and 3 seconds windows showed more spiky profile in all features which leads to the misclassification of normal and seizure EEG condition. On the other hand sliding windows of 4 seconds, and 5 seconds length have smooth amplitude profile of the features. A typical illustration of the spectral entropy profiles of 2 seconds, 3 seconds, 4 seconds and 5 seconds window lengths are discussed in following the paragraph.

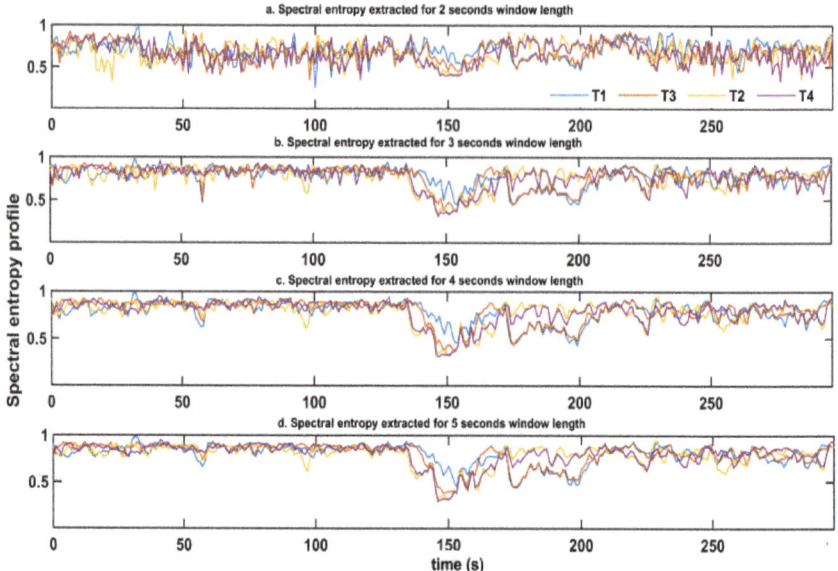

Figure 4.2: The spectral entropy profile of P1_s1 seizure obtained by sliding the window of 2 seconds, 3 seconds, 4 seconds and 5 seconds length over 5 minutes EEG is shown in (a), (b), (c), and (d) respectively.

The selection of optimal window length is illustrated using spectral entropy profile of P1_s1 seizure recording. The spectral entropy profile of T1, T3, T2 and T4 channels on different window lengths is shown in Fig. 4.2. Among the four window lengths, the short windows (2 seconds and 3 seconds) produce spiky patterns, and their features showing more variations in the profile. The larger windows (4 seconds and 5 seconds) produce smooth patterns which are more suitable to extract the characteristics of EEG. The same trend was noticed in the profile of all features extracted from the selected window lengths of EEG. Hence in this study, 4 seconds window sliding 1 second keeping 3 seconds overlap with previous window was selected to extract the features for the design of automated seizure detection system.

4.3 Significance of feature set

Each 5 minutes EEG recording was analyzed using 4 seconds sliding window with 3 seconds overlapping of the previous window. For each of 4 seconds long EEG data, seven time domain and fifteen frequency domain features were calculated. The following sections explain the analysis of significance of each feature in the automated detection of electrical seizure onset.

4.3.1 Time domain features

The time domain feature space consists of 7 features - mean, variance, skewness, kurtosis, signal energy, approximate entropy, and sample entropy of T1,T2,T3 and T4 channels. For all 29 seizure recordings, the profile of 7 features were analyzed. The features showing good separability of normal and seizure condition were selected as good descriptors for the automated seizure detection. In all 29 seizure recordings, it was observed that the mean computed for each 4 seconds window length EEG is zero in the normal state of the brain, and predominantly increased at the electrical onset of eleven seizures. Likewise, in twenty six seizure recordings, variance clearly captured the presence of amplitude increased rhythmic oscillations from the normal background EEG activity. The skewness showed noticeable increase at the electrical onset of P2_s4 seizure, and showed no consistent changes in other seizure recordings. In five seizures, kurtosis has large positive values which ensure the peaked distribution of seizure EEG samples, but showed no consistent changes in other recordings.

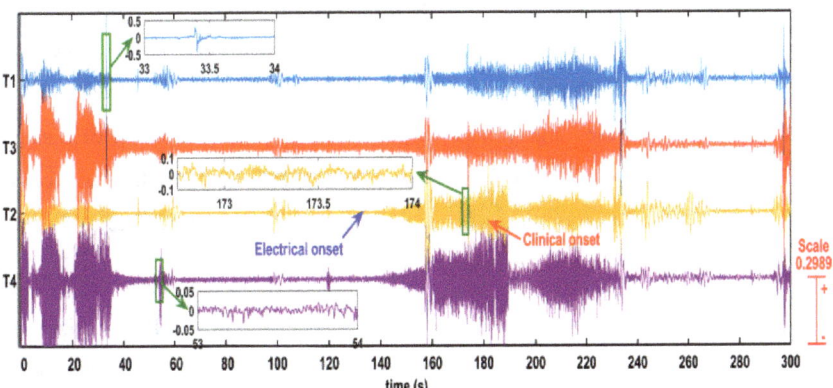

Figure 4.3: The filtered and normalized four channels EEG of P1_s1 seizure, 5 minutes recording with electrical and clinical onset at 138 seconds and 180 seconds respectively is illustrated.

The 5 minutes Scheme 1 normalized EEG of P1_s1 seizure, and its corresponding statistical measures is shown in Fig. 4.3 and Fig. 4.4 respectively. Comparing the

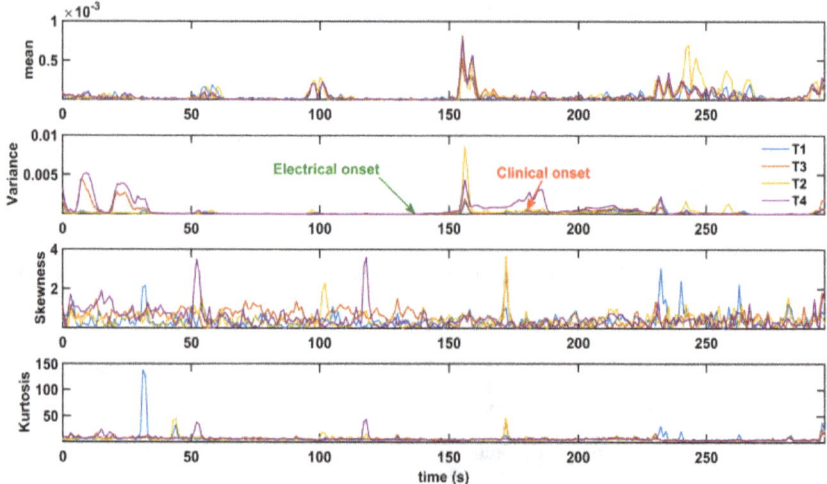

Figure 4.4: The statistical measures of P1_s1 seizure recording with electrical and clinical onset at 138 seconds and 180 seconds respectively is illustrated.

EEG with the statistical measures profile, the mean increases with the increase in EEG amplitude from the normal background activity. The variance measure showed significant changes, few seconds after the electrical onset of seizure, and the largest value is measured in T2 electrode covers the epileptic foci. The skewness, and kurtosis showed no consistent changes during the seizure activity. The skewness, and kurtosis features are more sensitive to large potentials due to non-seizure activity of the brain. The large increase in skewness around 53 seconds shown in Fig. 4.4 is because of sharp negative spiking highlighted in Fig. 4.3 inset representing T4 electrode potentials. The kurtosis increases largely at 33 seconds in T1 electrode pick-up for EEG samples with large potential difference from the back ground activity. Since the variance is equal to the energy of that zero mean signal, the signal energy profile of all 29 seizure recordings correlates with the variance profile of the recordings.

Similarly, 5 minutes EEG and the statistical features of P2_s4 seizure recording shows good variations in skewness and kurtosis for the electroencephalographic changes is illustrated in Fig. 4.5 and Fig. 4.6 respectively. The variance increases gradually at the electrical onset of seizure at 142 seconds and progress over the clinical onset at 180 seconds. The mean increases largely only few seconds after the clinical onset and is dominant at T2 electrode. The skewness and kurtosis showed increase in values during the seizure activity due to the spike and sharp wave discharges illustrated

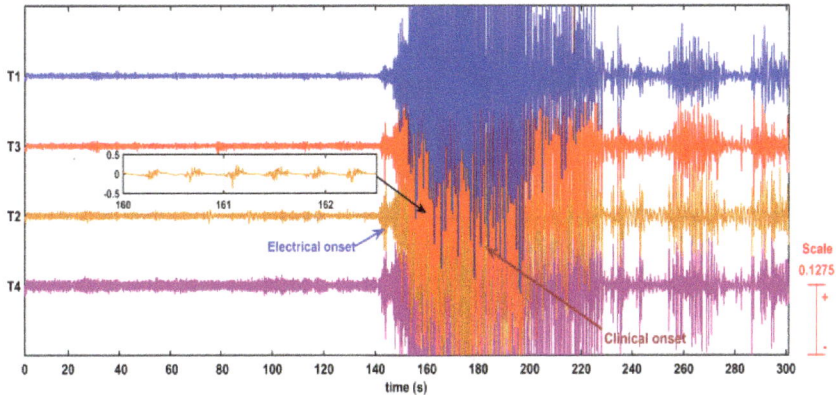

Figure 4.5: The 5 minutes EEG of P2_s4 seizure recording with electrical and clinical onset at 142 seconds and 180 seconds respectively is illustrated.

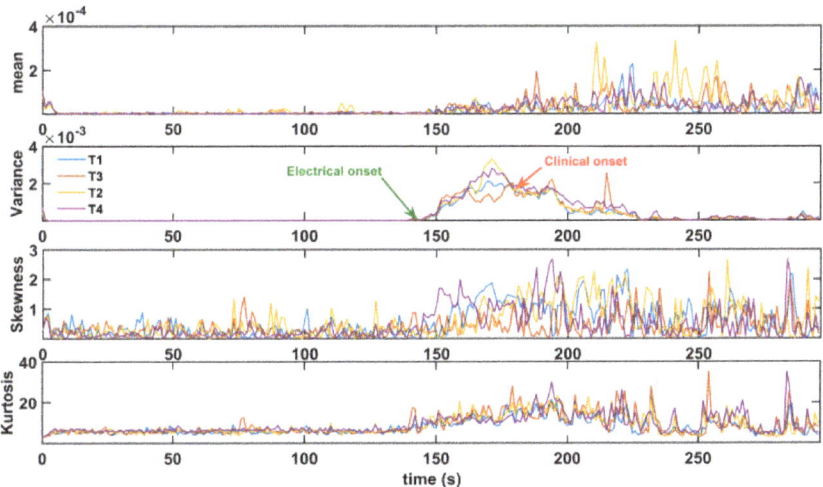

Figure 4.6: The statistical measures of P2_s4 seizure recording with electrical and clinical onset at 142 seconds and 180 seconds respectively is illustrated.

in Fig. 4.5 inset. Overall, the variance and signal energy features are good predictors of the electrical seizure onset. The other three features mean, skewness and kurtosis are positively increasing few seconds later the onset of seizure, and the presence of positive or negative spike and wave discharges is essential to identify the seizure using these three features.

In addition to signal energy and statistical measures to study the distribution, the approximate entropy (ApEn) and sample entropy (SampEn) were included to measure the irregularity in each 4 seconds windowed EEG signal. The ApEn and SampEn

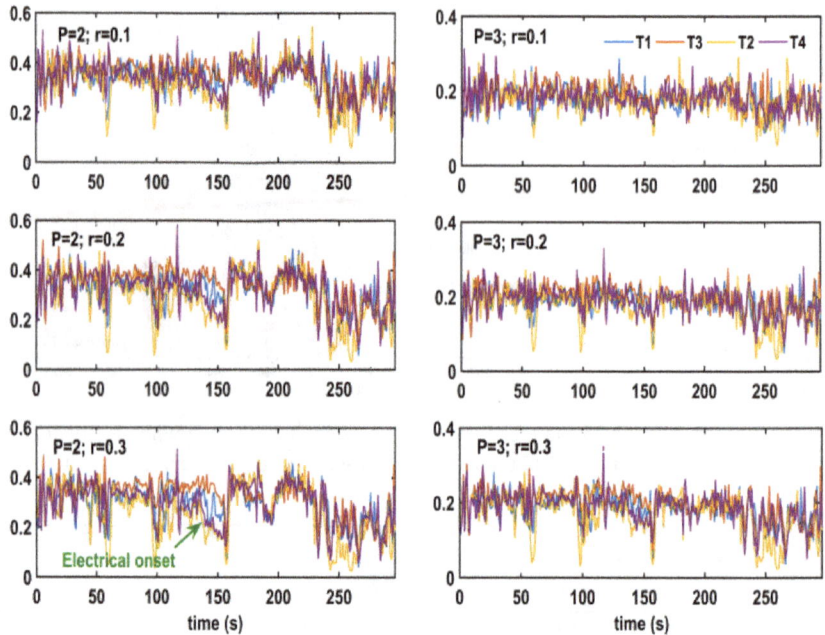

Figure 4.7: The profile of ApEn measures of P1_s1 seizure with electrical onset at 138 seconds in T2 electrode. The ApEn values are computed for p (2,3) and r (0.1,0.2,0.3) values for each 4 seconds windowed signal.

values were computed for a set of embedding dimension (p=2,3) and filtering threshold (r=0.1,0.2,0.3×standard deviation) values. After analysing all the 29 seizure recordings, it was noticed that the approximate entropy (ApEn) and sample entropy (SampEn) features showed increasing or decreasing trend at the electrical seizure onset. A typical example of ApEn and SampEn profile of P1_s1 seizure recording, for the selected p and r values (p=2; r=0.1, 0.2, 0.3) and (p=3; r= 0.1, 0.2, 0.3) for each 4 seconds (1600 samples) windowed signal is shown in Fig. 4.7 and Fig. 4.8 respectively.

In P1_s1 seizure, the ApEn values computed for embedding dimension (p=2), and filtering threshold (r=0.1,0.2, 0.3), decreases at the electrical onset of seizure. Specifically, the ApEn computed for p=2 and r=0.3 values, shows large decrease at the electrical onset at 138 seconds in T2 and T4 electrodes pick-up (Fig. 4.7). On the other hand, ApEn computed for p=3,r=0.3 parameters is not showing significant decrease at the seizure onset. Comparing the profile of statistical measures and ApEn measures as given in Fig. 4.4 and Fig. 4.7 respectively, the distribution of EEG samples – mean, skewness, and kurtosis influences the decrease in entropy values. Among

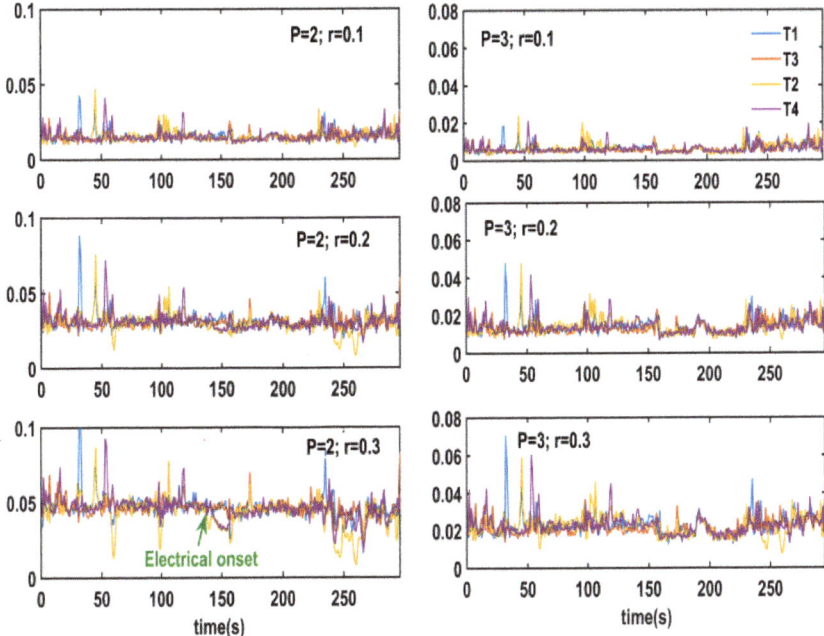

Figure 4.8: The profile of SampEn measures of P1_s1 seizure with electrical onset at 138 seconds in T2 electrode from right hemisphere brain. The SampEn values are computed for p (2,3) and r (0.1,0.2,0.3) parameters for each 4 seconds windowed signal.

the four statistical measures, the increased mean of time series around 60 seconds and 100 seconds influence the large decrease in ApEn values. The SampEn profile of 5 minutes P1_s1 seizure EEG is shown in Fig. 4.8. The SampEn computed for p=2 and r=0.3 shows sharp decrease at the electrical onset in T2 and T4 electrodes. Similarly, the SampEn computed for p=3 and r=0.3 values shows noticeable decrease at 159 seconds and the following seizure episode.

The 5 minutes EEG of P2_s4 seizure with electrical onset at 142 seconds in T2 electrode is illustrated in Fig. 4.9(B). The zoomed part of the normal EEG between 60 to 80 seconds is shown in Fig. 4.9(A). The normal EEG picked-up by T2, T4 electrodes were contaminated by high frequency noise whereas the T1, T3 electrodes have recorded noise-free potentials. The zoomed part of the seizure data between 136 seconds to 156 seconds is shown in Fig. 4.9(C). The seizure with 4 Hz single frequency oscillations starts at 142 seconds in right hemisphere was first recorded by T2, T4 electrodes. Later the EEG increases in amplitude and propagates into left

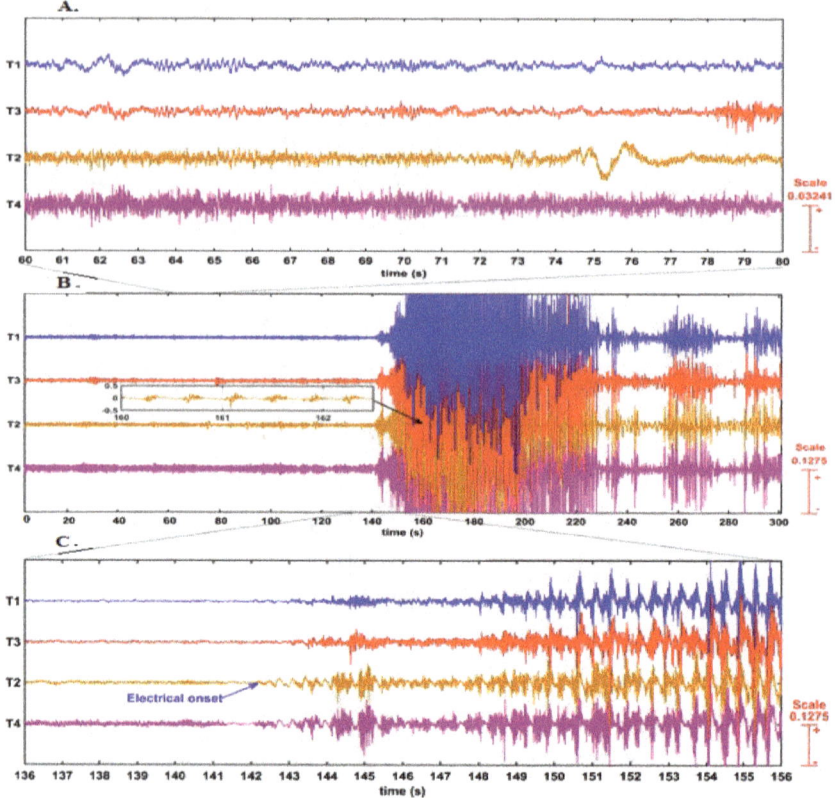

Figure 4.9: The normal and seizure condition of 5 minutes P2_s4 seizure EEG with electrical onset at 142 seconds in T2 electrode from right temporal region is shown in B. The zoomed part of the normal data between 60 to 80 seconds is shown in A. The zoomed part of seizure data between 136 seconds to 156 seconds is shown in C.

hemisphere brain is picked-up by T1, T3 electrodes. The corresponding ApEn and SampEn profiles of the 5 minutes EEG of P2_s4 seizure is illustrated in Fig. 4.10 and Fig. 4.11 respectively. Among the selected parameters, ApEn computed for p=0.2 and r=0.3 showed significant decrease at the electrical onset with single frequency oscillations in T2, T4 electrodes, and thereafter seizure progression as shown in Fig. 4.9(C). Comparing the 20 seconds (between 60 seconds and 80 seconds) normal EEG of P2_s4 shown in Fig. 4.9(A), the ApEn values of T1 and T3 electrodes are slightly lower than T2 and T4 electrodes which are obscured by artifacts. The ApEn profile is showing good separability of T1, T3 and T2, T4 electrodes in left and right hemisphere brain respectively. Contradictorily, the SampEn measures for P2_s4 showed

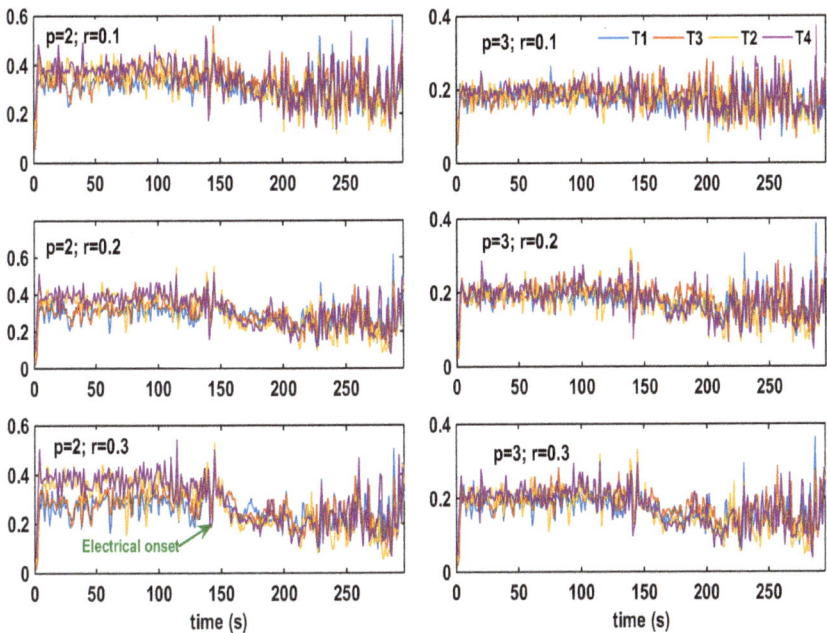

Figure 4.10: The profile of ApEn measures of P2_s4 seizure with electrical onset at 142 seconds in T2 electrode. The ApEn values are computed for p (2,3) and r (0.1,0.2,0.3) values for each 4 seconds windowed signal.

increasing trend at electrical onset of seizure, and the parameters selection p=3, and r=0.3 shows good separability among the left (T1,T3) and right (T2,T4) hemisphere electrodes with increasing values for T1, T3 electrodes. For all 29 seizure recordings, the ApEn and SampEn were computed for the selected p(2,3) and r(0.1, 0.2, 0.3) values. The profile of these features were carefully analysed at the electrical onset of seizures at the electrodes visually identified by the neurologists. The single frequency oscillations produced by the large synchronised neurons, decrease the irregularity and make the signal more deterministic in nature. It was expected that at the electrical onset of seizure, the ApEn and SampEn values should decrease to capture these rhythmic oscillations. Analysing all 29 seizure recordings, the ApEN and SampEn features showed noticeable decrease at the electrical onset for (p=2; r=0.3) and (p=3; r=0.3) values respectively for the 4 seconds (1600 samples) window length. Further the usefulness of these features were examined using univariate analysis method to study the distribution of normal and seizure feature samples.

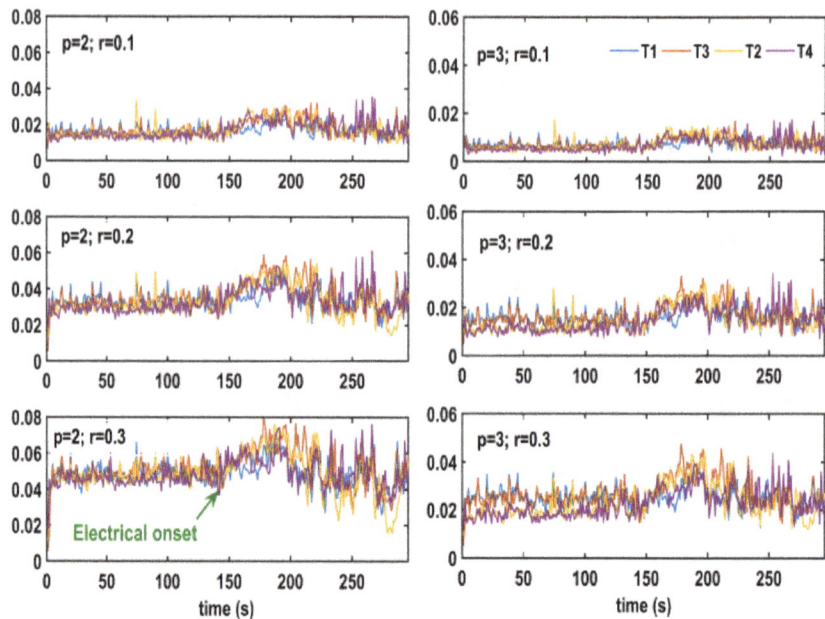

Figure 4.11: The profile of SampEn measures of P2_s4 seizure with electrical onset at 142 seconds in T2 electrode from right hemisphere brain. The SampEn values are computed for p (2,3) and r (0.1,0.2,0.3) parameters for each 4 seconds windowed signal.

Significance of time domain measures

Among the seven time domain features mean, variance, and signal energy profiles showed large increase at the electrical seizure onset. The other four measures skewness, kurtosis, approximate entropy and sample entropy were fluctuating over the course of time, and not showing consistent increase or decrease across the 29 seizure recordings.

The separability of normal features samples and seizure features samples by each time domain feature was examined by using univariate analysis (mean and quartile analysis) method (Fig. 4.12). The Box-plot technique is used to study the distribution of normal and seizure features samples with respect to median, first quartile (Q1), third quartile (Q3), minimum and maximum values of the samples distribution. The features samples are arranged in the order from minimum to the maximum value. The median is the middle of the ordered feature samples. The features samples sorted in descending order are divided into two halves with respect to the median. The middle of first half from minimum to median is named as 'first quartile (Q1)' and the middle of second half

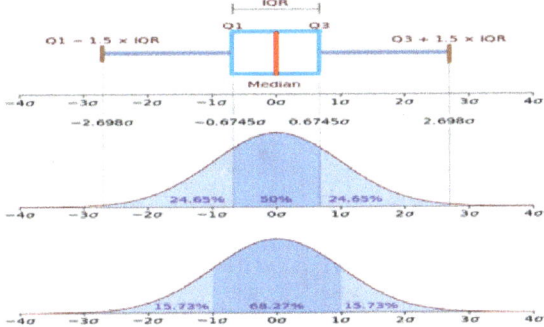

Figure 4.12: The sketch of the study of distribution of samples using Boxplot technique with Gaussian distribution from (Ord, 1972).

from median to maximum is named as 'third quartile (Q3)'. The inter quartile range (IQR) or the rectangular box shown in Fig. 4.12 is the measure of variability between 25th percentile (Q1) and 75th percentile (Q3) of the features samples which covers 50% of the features samples around the mean of distribution.

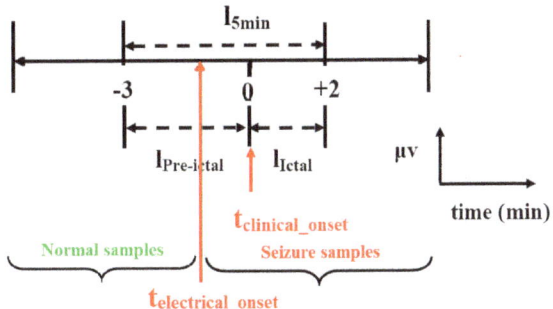

Figure 4.13: General schema of 5 minutes EEG data selection with respect to electrical onset of seizure for labeling the normal and seizure samples.

Seven time domain features were extracted for each 4 seconds window sliding over the 5 minutes EEG. The extracted features of all 29 seizures were grouped into normal and seizure (ictal) features samples with respect to the electrical seizure onset as illustrated in Fig. 4.13. The Box-plot of normal (green) and seizure (red) samples of all seven time domain features is shown in Fig. 4.14. The median (second quartile) of normal and seizure feature samples which very close and more overlap to the two regions are visible in mean, skewness, and kurtosis features.

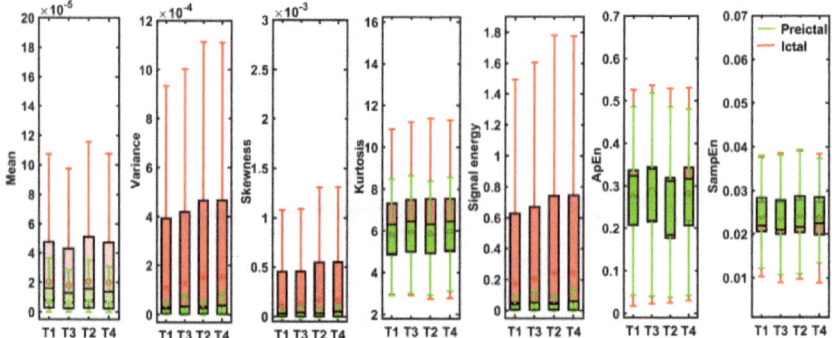

Figure 4.14: The Boxplot of normal and seizure features samples of all time domain measures.

Table 4.1: Range of values of seven features for normal and seizure EEG

Measures	Preictal EEG	Ictal EEG
Mean	$1.5e^{-05} \pm 4.3e^{-05}$	$4.2e^{-05} \pm 2.1e^{-04}$
Variance	$3.3e^{-05} \pm 2.0e^{-04}$	$3.6e^{-04} \pm 1.6e^{-03}$
Skewness	$1.6e^{-02} \pm 3.0e^{-01}$	$1.5e^{-02} \pm 3.1e^{-01}$
Kurtosis	5.80 ± 4.2	6.9 ± 10.0
Signal energy	$5.3e^{-02} \pm 3.1e^{-01}$	$5.7e^{-01} \pm 2.6$
ApEn	$2.6e^{-01} \pm 2.4e^{-01}$	$2.7e^{-01} \pm 2.7e^{-01}$
SampEn	$2.5e^{-02} \pm 1.5e^{-02}$	$2.4e^{-02} \pm 1.7e^{-02}$

Note: The range is defined as $\mu \pm 3\sigma$; The mean (μ) and standard deviation (σ) are the average of all the four channels T1,T2,T3 T4 of each measure.

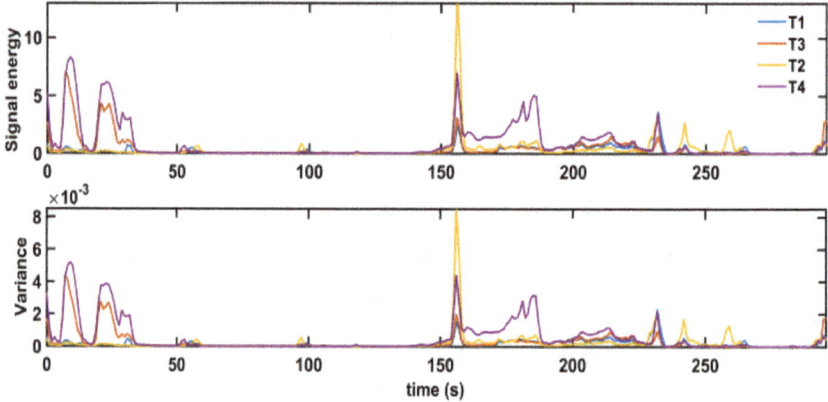

Figure 4.15: The signal energy and variance profile of P1_s1 seizure recording.

It was noticed that in all the four channels, the distribution of normal and seizure samples were completely overlapped in ApEn and SampEn features. Hence choosing mean, skewness, kurtosis, ApEn and SampEn features for the design of automated system will increase the chance for false positive (seizure) and missed seizure rate. The mean and quartile analysis of signal energy feature shows good separation of normal and seizure features samples about the order of ten. The normal and features samples of all the four channels of each feature were grouped. The mean and standard deviation of the normal and seizure samples of each feature were calculated. The range defined as the mean $\pm 3\sigma$ of normal and seizure features samples region were computed and tabulated in Table 4.1. Among the seven features, signal energy and variance showed good separation among the normal and seizure features samples. Since the variance and energy of the zero mean signal is same, the signal energy has been selected as the significant time domain feature to discriminate the normal and seizure EEG (Fig. 4.15).

4.3.2 Frequency domain measures

The EEG patterns of seizure onset appear in different forms varying in amplitude, and frequency including electrodecremental, rhythmically evolving delta, theta, and alpha frequencies, rhythmic spiking and spike-wave discharges (Ebersole and Pacia, 1996; Fisher *et al.*, 2014). During the seizure, the hyper synchronised neurons produce rhythmic oscillations or rhythmic spike discharges at different frequencies. In this work, spectral energy and wavelet energy were extracted in 1 to 3, 3 to 6, 6 to 12, 12 to 25, 25 to 50, 50 to 100, and 100 to 200 Hz frequency bands to capture these rhythmic patterns.

These frequency domain features were extracted for each 4 seconds window sliding over the 5 minutes EEG. For all 29 seizures, the profile of these features were analyzed with respect to the electrical and clinical seizure onset identified by the Neurologists. In all 29 seizure recordings, it was observed that the spectral (E1, E2, E3, E4) and wavelet (Ea6, Ed6, Ed5, Ed4) energies in 1 to 3 Hz, 3 to 6 Hz, 6 to 12 Hz and 12 to 25 Hz frequency bands showed increasing trend during the seizure episode. The 5 minutes EEG of P1_s1 seizure and its spectral energy and wavelet energy profiles are shown in Fig. 3.7, Fig. 4.16 and Fig. 4.17 respectively. The spectral energy (E2) and wavelet energy (Ed6) in 3 to 6 Hz band increases predominantly at the electrical seizure onset at 138 seconds in T2 and T4 electrodes. Later, the seizure activity is captured by energy

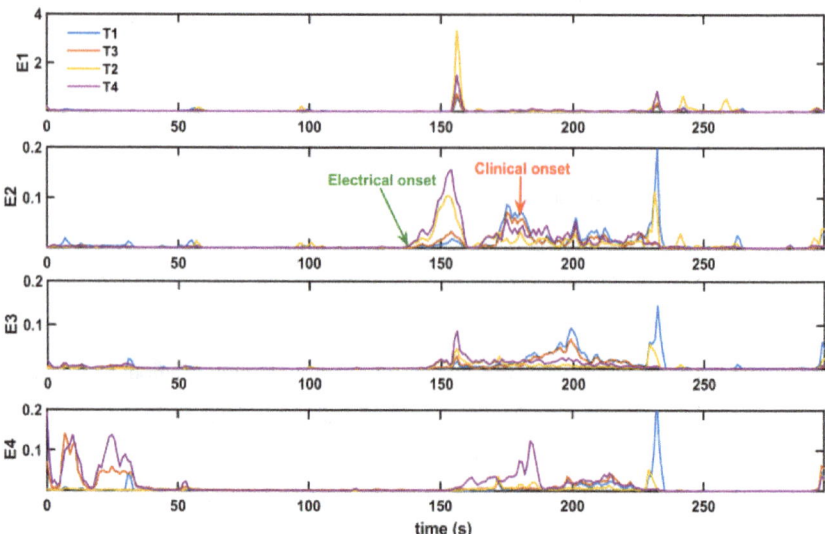

Figure 4.16: The spectral energy profiles of seizure P1_s1 starts at 138 seconds from right hemisphere is picked up by T2, T4 electrodes. The spectral energy measured in 1 to 3 Hz (E1), 3 to 6 Hz (E2), 6 to 12 Hz (E3), 12 to 25 Hz (E4) frequency bands is shown.

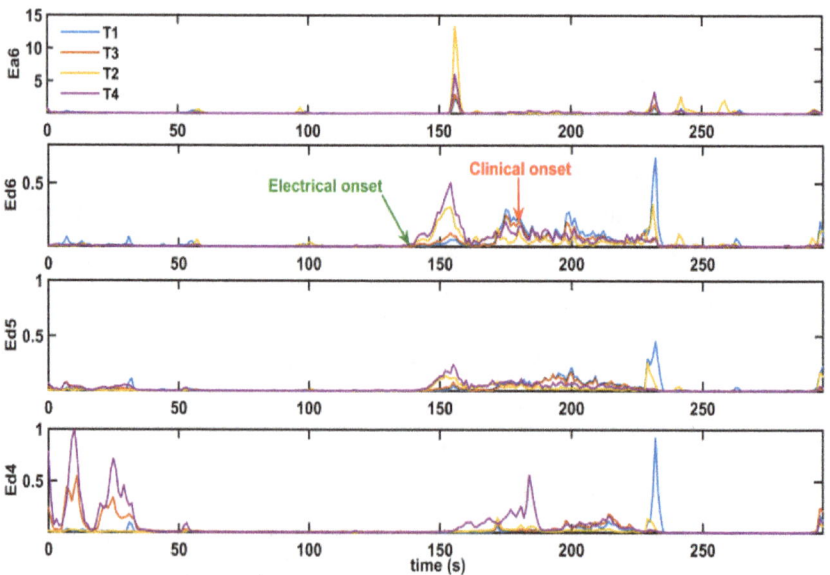

Figure 4.17: The wavelet energy profiles of P1_s1 seizure originates at 138 seconds from right hemisphere is picked up by T2, T4 electrodes. The wavelet energy measured in 1 to 3 Hz (Ea6), 3 to 6 Hz (Ed6), 6 to 12 Hz (Ed5), 12 to 25 Hz (Ed4) frequency bands is shown.

(E3,Ed5 and E4,Ed4) rise in 6 to 12 Hz and 12 to 25 Hz bands respectively. The large voltage appears in seizure EEG at 158 seconds which is more prominent in T2 electrode increase the band specific energy (Ea6, E1) in 1 to 3 Hz band.

In all 29 seizure recordings, the profile of wavelet energy is correlated with spectral energy of all the frequency bands. The decomposition of EEG signal using discrete wavelet transform (DWT) is mainly depending on the sampling frequency used for recording the EEG. Comparatively, the spectral energy computation is more simple, having flexibility to select the frequency band width of interest. Hence the simple and efficient spectral energy features were selected for the design of automated system to detect the electrical seizure onset. The spectral energy in 1 to 3, 3 to 6, 6 to 12, and 12 to 25 Hz bands showed increasing trend at electrical onset of all 29 seizures. Hence these low frequency were combined, the entire signal was divided into 1 to 25 Hz ($E_{lowband}$), 25 to 100 Hz ($E_{midband}$), and 100 to 200 Hz ($E_{highband}$) bands for the measurement of spectral energy. The spectral entropy feature was extracted to measure the regularity within the 4 seconds EEG in specific frequency band. Particularly, the presence of synchronous activity within 3 to 12 Hz was captured by the spectral entropy feature computed in 3 to 12 Hz band.

4.3.3 Significance of features - Quantitative analysis

The significance of features set – mean, variance, skewness, kurtosis, signal energy, approximate entropy, sample entropy, spectral entropy, $E_{lowband}$, $E_{midband}$, and $E_{highband}$ extracted from T1, T3, T2 and T4 channels (11 features x 4 channels = 44 features) were quantitatively analyzed using ReliefF, a filter based features selection algorithm. The filter based method selects the significant features independent of classifiers performance. The ReliefF is a K-nearest neighbours based feature selection algorithm, to evaluate the importance of features used for training the classifiers (Kononenko, 1994; Kononenko et al., 1997). All the 44 features' normal and seizure samples of all 29 seizure recordings with '0' and '1' response variables respectively were passed through the ReliefF algorithm. The reliefF algorithm evaluated the relevance of the 44 features and rank them according to the estimated feature significance weights as shown in Fig. 4.18. The signal energy extracted from all the four channels scored highest significance weights and top the features list. Following the signal energy features, the $E_{lowband}$ features are more significant in distinguishing the normal EEG and seizure condition. The $E_{midband}$, $E_{highband}$ and Spectral entropy in 3 to 12 Hz features scored more weights and ranked within top 20 positions. The statistical features mean, variance, skewness and kurtosis are not scoring more importance to discriminate the normal and seizure samples. The ApEn and SampEn features scored very low feature importance weights and are not suitable for efficiently classifying the normal and seizure features samples. Among the 11 features, the signal energy, spectral energy in 1 to 25 Hz, 25 to 100 Hz and 100 to 200 Hz bands, and spectral entropy in 3 to 12 Hz band ranked the top 20 positions. Hence these five features were considered for the design of seizure detection system.

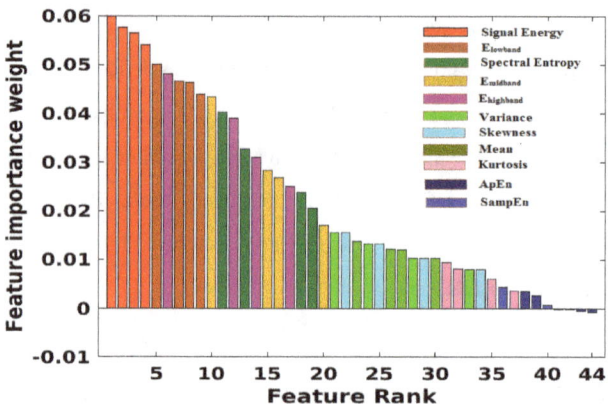

Figure 4.18: The significance of features computed using ReliefF a filter based feature selection algorithm.

4.4 Reduced feature set

As discussed in Section. 4.3, based on the visual and quantitative analysis, among the seven time domain features – mean, variance, skewness, kurtosis, signal energy, approximate entropy and sample entropy, the signal energy was selected as the best time domain feature to discriminate the normal and seizure condition. The spectral entropy, spectral energies in $E_{lowband}$, $E_{midband}$, and $E_{highband}$ features were selected in the frequency domain.

From the 29 seizure recordings, it was noticed that the spectral energy in $E_{lowband}$ increased the most at electrical seizure onset, whereas $E_{midband}$ and $E_{highband}$ increased few seconds later the seizure onset. The spectral energy in high frequency bands 25 to 100 Hz and 100 to 200 Hz increased only during the later part of ictal cycle, and their energy level is quite lower than the energy in 1 to 25 Hz frequency band. Moreover, compared to T1 and T2 electrodes, the spectral energy in 25 to 100 Hz and 100 to 200 Hz bands showed large variations in T3 and T4 electrodes recording. The five features extracted from 5 minutes EEG of P1_s1 seizure recording are shown in Fig. 4.19. The spectral entropy varied most at electrical onset and contributed to the detection of the electrical onset of seizure. The $E_{lowband}$ spectral energy and signal energy remained the same at the electrical onset, and became large only few seconds after the electrical onset of seizure.

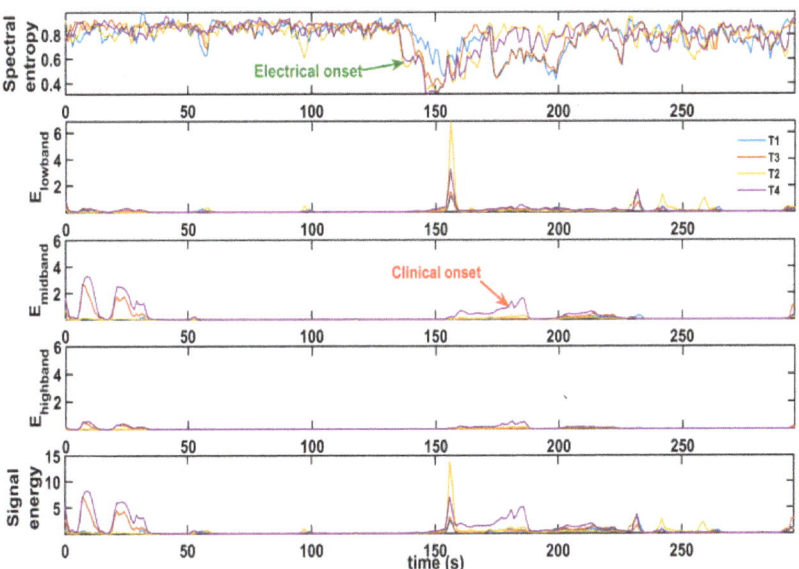

Figure 4.19: Prominent features extracted from the P1_s1 seizure recording normalized using Scheme 1. The spectral entropy decreased largely at the electrical onset at 138 seconds and contributed most to the detection of electrical onset.

Based on the quantitative and visual analysis, the features showing large changes from the background activity – spectral entropy in 3 Hz to 12 Hz, $E_{lowband}$ spectral energy in 1 to 25 Hz band, and signal energy in 1 to 200 Hz – were selected as best candidates for the automated seizure onset detection system. The spectral entropy in 3 to 12 Hz band significantly decreased at the electrical onset of 26/29 recorded seizures. In 27/29 recorded seizures, the selected three features showed consistent increase or decrease at their electrical onset. The two seizures P2_s1 and P7_s3 showed distinctive changes in all three features only at the clinical seizure onset. In TLE, spread of electrical activity from temporal lobe to neighboring areas results in loss of consciousness and tonic-clonic activity. The spectral entropy in 3 to 12 Hz was selected to capture the rhythmic spiking or oscillations of large synchronized neuronal populations at the electrical onset of a seizure. The spectral energy in 1 to 25 Hz frequency band was selected to characterize the EEG dynamics at the initial stage of a seizure episode with loss of consciousness. The signal energy feature was selected to capture large energy changes at different stages of the seizure episode. The features – spectral entropy in 3 to 12 Hz, spectral energy in 1 to 25 Hz, and signal energy in 1 to 200 Hz yielding better distinguishability of normal and seizure samples were selected to train the classifiers to perform the automated electrical onset detection.

4.5 Patient independent system

An automated seizure detection system was designed by using the five supervised learning algorithms namely Linear Discriminant Analysis (LDA), Naive Bayes (NB), Decision Tree (DT), Support Vector Machine (SVM), and K Nearest Neighbor (KNN). The LDA classifier predicts the class of the unknown features sample using the estimated mean and variance of the training features for each class. It assigns the class label for unknown features sample according to the estimated probability that the features sample belongs to each class. The NB is a probabilistic classifier based on Bayes theorem. It assumes that the input features samples of a class are unrelated to each other. The DT algorithm is a structured tree which defines the set of rules from training features samples of each class. Based on the outcome of the rule, the features sample with unknown label is classified. The KNN is a non-parametric classifier based on memory, does not require a model to fit. The unknown features sample is assigned to the class most common amongst its K nearest neighbors. In this work the unknown features samples were assigned to its K=2 nearest neighbors. The SVM classifier assigns the label to the unknown features samples according to the discriminant function defined by a hyperplane. The hyperplane which maximize the separation between the classes are viewed in high-dimensional space. In this work, SVM with Radial Basis Function (RBF) kernel was used to design the seizure detection system. The grid search algorithm was used to optimize the SVM parameters such as box-constraint and kernel scale of the SVM classifier. The detailed description of each machine learning algorithm is given in Appendix A.

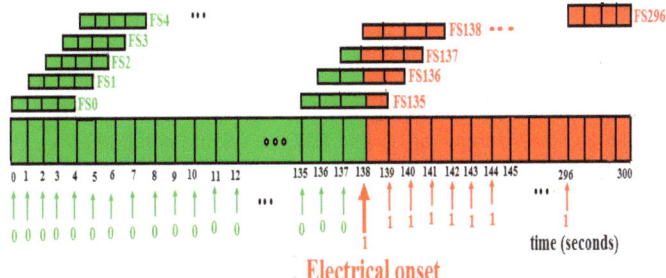

Figure 4.20: Schematic of the labeling of normal and seizure features samples of P1_s1 EEG with respect to the electrical onset at 138 seconds visually identified by the Neurologists.

These classifiers were trained to solve the binary classification problem of distinguishing the difference between the normal and seizure classes. The classifiers were trained using three features – spectral entropy, $E_{lowband}$, and signal energy of the 4 channels T1,T2,T3 and T4 to distinguish the normal and seizure condition. The 12 features set (4 channels x 3 features) of each 5 minutes recording was split into normal and seizure features samples with respect to the electrical seizure onset (Fig. 4.13). A typical example of normal and seizure features samples selection of P1_s1 recording is illustrated in Fig. 4.20. The electrical onset of P1_s1 seizure was visually identified by the Neurologists at 138 seconds as shown in Fig. 3.7. The 12 features were extracted for the EEG of 4 seconds window sliding over the length of 300 seconds EEG. For each 4 seconds window starting from time=0 seconds to time=296 seconds, 297 features samples (FS0 to FS296) were extracted. Since the length of EEG at time=297 seconds is less than 4 seconds long, the last 3 seconds EEG was discarded. Among the set of 297 features samples (297 x 12), 138 features samples (from FS0 to FS137) were labeled as zero and the remaining 159 features samples from FS138 to FS296 were labeled as one to represent the features samples extracted from normal and seizure EEG respectively. Similarly, the normal and seizure features samples of $N-1$ patients' seizures recordings were grouped and labeled as zeros and ones respectively to model the classifiers. The features samples beyond the $\pm 3\sigma$ (where σ is the standard deviation) limit were discarded. This was done to prevent the noise from being tampered with the classification if any over fitting is encountered. This is also known as the six-sigma rule to eliminate the outliers, and to generate an appropriate data set to learn the classification algorithm. If any one feature in the features sample exceeds the $\pm 3\sigma$, the entire features sample (1x12 features) was rejected in training the classifier. The features samples extracted from N^{th} patient's seizure recordings were passed through the trained classifier model. For every 1 second, the classifier identify the given unknown features sample as normal EEG or seizure condition according to the given 1x12 features samples. Thus the classifier performance was validated using an unknown patient data which ensures the patient independent system approach (Chua *et al.*, 2011; Orosco *et al.*, 2016). The entire 5 minutes EEG of each seizure recording was tested with the set of five classifiers. The performance of each classifiers was evaluated using the performance measures given in the Eq. (4.17).

$$Sensitivity = \frac{No.\ of\ correctly\ classified\ seizure\ samples}{Total\ no.\ of\ seizure\ samples} \times 100\ \%$$

$$Specificity = \frac{No.\ of\ correctly\ classified\ normal\ samples}{Total\ no.\ of\ normal\ samples} \times 100\ \%$$

$$Accuracy = \frac{No.\ of\ correctly\ classified\ samples}{Total\ no.\ of\ normal\ and\ seizure\ samples} \times 100\ \%$$

$$Error\ rate = \frac{No.\ of\ misclassified\ samples}{Total\ no.\ of\ normal\ and\ seizure\ samples} \times 100\ \%$$

$$False\ Positive\ Alarms = 1 - (Specificity/100)$$

(4.17)

The *Sensitivity* is defined as the ability of classifier to classify the seizure samples accurately, whereas the *Specificity* is defined as the ability of classifier to correctly classify the normal samples. The *False Positive Alarms (FPA)* is defined as the number of normal samples classified as seizure samples (False seizure detections). The *Accuracy* is defined as the classifier's ability to correctly classify the normal and seizure samples. The *Error rate* is the classifier's performance measure to define the number of misclassified samples among the given samples. For each classifier, there is a trade-off between sensitivity and specificity. The detection performance of each classifier was validated on patient independent data set. The sensitivity and specificity measures of five classifiers distinguishing the normal, and seizure features samples of each seizure recording were compared. The classifiers were trained by using two set of features. The classifiers were trained by using only time domain features – signal energy of four channels, and studied the performance of classifiers to detect the electrical onset of seizures. Later, the classifiers were trained by using all three significant features – spectral entropy, $E_{lowband}$ spectral energy and signal energy features of four channels, and the improvement in performance of classifiers to detect the electrical onset of seizures were investigated. Primarily, the significance of signal energy features on automated seizure detection was investigated and the detection performance is presented in Section. 4.5.1. The classifiers were trained and tested using the signal energy features of T1, T2, T3 and T4 channels as per the data selection procedure described in the earlier part of this section. The performance of each classifier in detecting the normal and seizure samples of each seizure recording was analyzed. The detection latency (which is defined as the time difference between the visual and automated detection of electrical onset of seizure) of each classifier on individual seizure recording was described. Likewise, the performance of classifiers trained using all three features is explained in Section. 4.5.2. The detection latency and the performance of five classifiers trained using all three features were compared with the classifiers trained using signal energy features alone.

4.5.1 Training of classifiers using signal energy features

The five classifiers were trained by using the signal energy features extracted from T1, T2, T3, and T4 channels for the automated detection of electrical seizure onset. The trained classifiers labeled the given unknown features samples as normal or seizure condition according to the given (1x4) signal energy features. As mentioned in Section. 3.1.1, the length of five seizure recordings namely P7_s2, P7_s3, P8_s1, P8_s3 and P12_s1 had less than 5 minutes EEG due to the removal of EEG severely contaminated by artifacts during the clinical seizure. Using patient independent approach (leave one patient approach) the LDA, NB, DT, SVM and KNN classifiers were tested on all 29 seizure recordings. According to the given 1x4 features sample, for every 1 second, the classifier identify the unknown features sample as normal EEG or seizure condition. A typical example of the classification of normal and seizure condition of P1_s1 recording by SVM classifier trained using signal energy features is shown in Fig. 4.21. As shown in Fig. 4.20, P1_s1 recording consists of 138 normal features samples and 159 seizure features samples represented using signal energy of T1, T3, T2 and T4 channels. The normal features samples are represented as '0' and the seizure features samples are represented as '1'. The SVM classifier trained using signal energy features identified the normal features samples with 77% of specificity, and the seizure feature samples were identified with 66% of sensitivity. The misclassification of normal and seizure features samples are marked as 'False seizure detection' and 'False normal detection', respectively. Comparing the signal energy profile and SVM output, it was noticed that the increase in signal energy at the beginning of seizure recording produced more false seizure detections. The SVM classifier continuously identified the given seizure features samples starting from 150 seconds and the automated seizure onset detection was marked at 150 seconds.

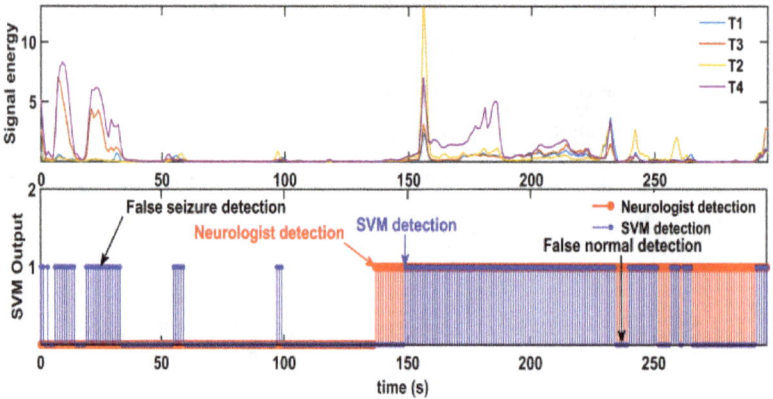

Figure 4.21: The classification of normal and seizure condition of P1_s1 recording by SVM classifier trained using signal energy features (Scheme 1).

The SVM classifier detected the onset of P1_s1 seizure at 150 seconds which is 12 seconds later the visual detection (138 seconds) by the Neurologists. Similarly, the detection performance of other four classifiers were tested on 29 seizure recordings to detect the normal features samples and seizure features samples correctly. Using patient independent approach, the performance of classifiers tested on each seizure recordings is tabulated in Table 4.2. Comparing the detection performance of five classifiers, LDA and NB classifiers scored high specificity in identifying the given features samples as normal EEG. On the other hand, these two classifiers scored low sensitivity to identify the given features samples as seizure condition. Among the 29 seizure recordings, LDA classifier had very low sensitivity for P1_s2, P18_s1 and P19_s1 seizures by identifying few seizure features samples after their clinical seizure onset. Likewise, NB classifier identified few seizure features samples of P18_s1 seizure recording after the clinical seizure onset. On the other hand, NB classifier had very low sensitivity for P1_s2 seizure by detecting few seizure features samples at the beginning of seizure episode.

The other three classifiers DT, SVM and KNN, had good sensitivity performance measures on 29 seizure recordings. Among the three classifiers, SVM classifier had good specificity in identifying the normal features samples. The DT classifier showed more false seizure detection (low specificity) for P12_s1, P15_s1, P16_s1 and P17_s1 seizure recordings. Similarly, the KNN classifier showed more false seizure detection for P12_s1, P16_s1 and P19_s1 seizure recordings. Among the 29 seizures, NB, DT, KNN and SVM classifiers scored very low specificity in identifying the normal features samples of P4_s1 and P15_s1 seizure recordings. Particularly, the SVM classifier classified all the normal samples of P4_s1 recording as seizure samples.

The classifiers identified the given normal features samples and seizures features samples of each seizure recording with different specificity and sensitivity values respectively. The average performance measures of each classifier trained using signal energy features, validated on 29 seizure recordings using patient independent approach is given in Table 4.3. Among the five classifiers, LDA classifier had the highest specificity of 95% with the lowest sensitivity of 44%. The NB classifier had the specificity and sensitivity of 90% and 54% respectively. The KNN classifier scored 74% and 67% of sensitivity and specificity respectively with the error rate of 30%. The DT classifier scored the average sensitivity and average specificity of 75% and 72% with the error rate of 27%. The SVM classifier scored the misclassification rate of 27% with the average sensitivity and average specificity of 64% and 82% respectively.

The automated detection of electrical onset of each seizure was marked at the time from which the classifiers able to continuously detect the given seizure features samples correctly. The automated detection of electrical onset of 29 seizures by all the five classifiers were compared with their visual detection by the Neurologists. The visual and automated detection of electrical onset of 29 seizures by all the five classifiers trained using signal energy features are presented in Table 4.4. Comparing the visual

Table 4.2: The performance of classifiers for individual seizures with signal energy measures (Scheme 1)

Seizure ID	Sensitivity (%)					Specificity (%)				
	LDA	NB	DT	SVM	KNN	LDA	NB	DT	SVM	KNN
P1_s1	33	48	72	66	69	83	81	74	77	66
P1_s2	03	14	33	39	28	100	99	97	99	67
P1_s3	80	85	92	92	87	99	99	81	98	66
P2_s1	33	38	60	56	59	100	100	97	99	94
P2_s2	48	53	65	57	65	99	99	88	99	61
P2_s4	46	46	68	54	66	100	100	99	99	43
P3_s1	31	40	96	48	95	100	99	97	99	99
P4_s1	85	97	97	61	97	68	15	23	0	38
P5_s1	48	56	81	59	80	96	95	79	93	82
P7_s1	37	42	97	45	100	98	98	93	96	99
P7_s2	44	45	80	47	73	99	99	88	99	81
P7_s3	55	58	72	60	73	100	100	95	100	77
P8_s1	82	87	95	100	89	95	93	70	88	61
P8_s3	46	55	72	74	68	99	99	86	98	66
P11_s1	27	36	68	43	70	100	99	96	98	96
P11_s2	24	32	67	39	69	100	99	96	98	96
P11_s3	35	40	93	43	98	98	98	93	96	99
P12_s1	93	98	100	100	99	70	60	17	36	19
P13_s1	46	58	74	71	71	98	98	69	97	61
P14_s1	53	85	47	78	69	85	57	78	58	83
P15_s1	77	77	97	88	94	94	81	10	09	36
P16_s1	70	78	96	90	87	94	86	14	70	09
P17_s1	59	68	96	93	93	99	98	31	86	45
P17_s2	34	46	73	69	73	99	99	55	96	78
P18_s1	17	25	25	34	26	99	98	98	96	85
P19_s1	13	39	71	79	78	95	88	75	87	30
P20_s1	51	62	93	77	87	99	99	87	97	86
P20_s2	60	68	86	98	88	96	91	52	66	44
P21_s1	31	61	72	55	57	75	92	49	39	55

Note: Pn_sm - n^{th} patient m^{th} seizure recording; LDA-Linear Discriminant Analysis; NB-Naive Bayes ; DT-Decision Tree; SVM-Support Vector Machine; KNN-K Nearest Neighbour; Sensitivity=TS/(TS+FN); Specificity=TN/(TN+FS); TS-True Seizure; TN-True Normal; FS-False Seizure; FN-False Normal

Table 4.3: The average performance of classifiers trained using signal energy (Scheme 1)

Classifier / Parameter	LDA	NB	DT	SVM	KNN
Sensitivity(%)	44	54	75	64	74
Specificity(%)	95	90	72	82	67
Accuracy(%)	70	72	73	73	70
Error Rate(%)	30	28	27	27	30

Note: Para - Parameters; Clas - Classifiers; Sensitivity=TS/(TS+FN); Specificity=TN/(TN+FS); Accuracy=(TS+TN)/(Total no. of samples); Error rate=1-accuracy; TS-True Seizure; TN-True Normal; FS-False Seizure; FN-False Normal

Table 4.4: Detection of electrical seizure onset using signal energy measures (Scheme 1)

Seizure ID	Visual detection (s)		Automated detection (s)								
	EO	CO	LDA	LDA Latency	NB	NB Latency	DT	DT Latency	SVM	SVM Latency	KNN	KNN Latency
P1_s1	138	180	156	18	156	18	141	3	150	12	150	12
P1_s2	143	180	ND	-	163	20	148	5	162	19	149	6
P1_s3	165	180	169	4	169	4	162	-3	168	3	168	3
P2_s1	174	180	195	-	194	-	192	-	192	-	191	-
P2_s2	159	180	175	16	175	16	175	16	172	13	171	12
P2_s4	142	180	153	11	153	11	152	10	148	6	147	5
P3_s1	152	180	220	-	220	-	153	1	220	-	151	-
P4_s1	159	180	154	-5	MS	-	MS	-	MS	-	MS	-
P5_s1	165	180	168	3	168	3	167	2	167	2	167	2
P7_s1	175	180	258	-	253	-	180	-	243	-	174	-1
P7_s2	157	180	188	-	187	-	179	22	186	-	187	-
P7_s3	173	180	220	-	219	-	220	-	216	-	206	-
P8_s1	177	180	177	0	174	-3	169	-8	173	-4	169	-8
P8_s3	121	180	183	-	182	-	152	31	151	30	148	27
P11_s1	134	180	220	-	219	-	205	-	204	-	203	-
P11_s2	114	180	200	-	199	-	185	-	184	-	184	-
P11_s3	176	180	261	-	256	-	179	3	246	-	178	2
P12_s1	162	180	165	3	164	2	MS	-	MS	-	MS	-
P13_s1	161	180	181	-	181	-	173	12	176	15	176	15
P14_s1	125	180	134	9	127	2	131	1	126	1	128	3
P15_s1	145	180	146	1	146	1	MS	-	MS	-	MS	-
P16_s1	163	180	166	3	166	3	MS	-	161	-2	MS	-
P17_s1	163	180	197	-	182	-	164	1	171	8	164	1
P17_s2	138	180	207	-	205	-	160	22	152	14	201	-
P18_s1	141	180	ND	-	ND	-	158	17	157	16	155	14
P19_s1	114	180	ND	-	136	22	166	52	131	17	MS	-
P20_s1	134	180	150	16	140	6	140	6	137	3	147	13
P20_s2	167	180	170	3	169	2	168	1	166	1	165	-2
P21_s1	111	180	157	46	145	34	119	8	127	16	153	42
Mean				9		9		10		10		8.3
Median				3.5		4		5.5		12		4
Minimum				-5		-3		-8		-4		-8
Maximum				46		34		52		30		42

Note: Pn_sm - n^{th} patient m^{th} seizure recording; ND - Not Detected - the seizure feature samples only at the end of seizure episode were classified correctly; MS - More number of normal samples were classified as Seizure condition; AS - All the normal samples were classified as Seizure condition

and automated detection of electrical onset of each seizure, the detection latency was computed as defined in Section. 4.5 and reported in Table 4.4. The positive sign of latency represents the automated detection of electrical onset happens after the visual detection of onset, and negative sign represents the automated detection before the visual detection by the Neurologists. When only considering the seizures detected before their clinical onset, the LDA classifier identified the electrical onset of fourteen seizures with average latency of 9 seconds (ranging from -5 seconds to 46 seconds with the median of 3.5 seconds). The NB classifier identified the electrical onset of fifteen seizures with average latency of 9 seconds (ranging from -3 seconds to 34 seconds with the median of 4 seconds). The DT classifier identified the electrical onset of twenty seizures with average latency of 10 seconds (ranging from -8 seconds to 52 seconds with the median of 5.5 seconds). The KNN classifier identified the electrical onset of eighteen seizures with average latency of 8.3 seconds (ranging from -8 seconds to 42 seconds with the median of 4 seconds).

The SVM classifier identified the electrical onset of eighteen seizures with average latency of 10 seconds (ranging from -4 seconds to 30 seconds with the median 12 seconds). The classifier suitable for automated seizure detection system could be able to detect the onset of seizure with very good specificity measures. The automated detection system with high specificity measures reduce the false alarms causing the patients more depressed on expecting a seizure. The SVM classifier having good specificity and sensitivity performance measures is more suitable for the design of electrical seizure onset detection system.

4.5.2 Training of classifiers using all three features

The classifiers trained using signal energy features in Section. 4.5.1, scored large latency in detecting the electrical seizure onset. For reducing the detection latency, the spectral entropy in 3 to 12 Hz and spectral energy in 1 to 25 Hz were included to pick-up the earliest changes in EEG at the electrical onset of seizure. The five classifiers were trained by using spectral entropy in 3 to 12 Hz, $E_{lowband}$ in 1 to 25 Hz, and signal energy in 1 to 200 Hz of T1,T2,T3 and T4 channels (3 x 4 = 12) features extracted from EEG normalized using seizure specific maximum and minimum values. As explained in Section. 4.5, the normal and seizure features samples of $N-1$ patients' seizures recordings were grouped and labeled as zeros and ones respectively to train the classifiers. The features samples extracted from N^{th} patient's seizure recordings were used to test the trained classifiers. Using patient independent approach (leave one patient approach) classifiers were tested on 29 seizure recordings. According to the given 1x12 features sample, for every 1 second, the classifier identify the unknown features sample as normal EEG or seizure condition. For an example, the spectral entropy, $E_{lowband}$ spectral energy and signal energy features extracted from P2_s1 seizure recording and their corresponding SVM output for automated detection of normal and seizure features samples is given in Fig. 4.22.

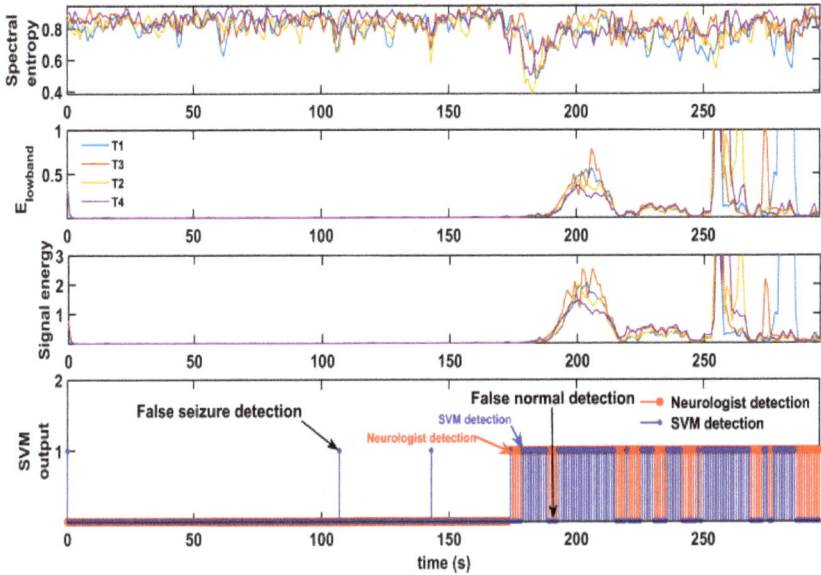

Figure 4.22: The significant features extracted from P2_s1 seizure normalized using channel specific maximum and minimum values. The seizure electrical onset at 174 seconds from right temporal lobe T2 electrode. The classification of normal and seizure condition of P2_s1 seizure by SVM classifier trained using three features.

The electrical onset of P2_s1 seizure was visually identified by the Neurologist at 174 seconds from right temporal lobe at T2 electrode as shown in Fig. 3.14. The 4 seconds window sliding over the 5 minutes EEG of P2_s1 seizure recording extracted 297 features samples (FS0 to FS296). The P2_s1 seizure recording consists of 174 normal features samples and 123 seizure features samples. The SVM classifier trained using three features had the specificity (sensitivity) of 98% (63%) to identify the normal (seizure) features samples (Table 4.5). The spectral entropy showed gradual decrease starting from the time of electrical onset of seizure picked up by T2 and T4 electrodes from the right temporal region (Fig. 4.22). When the spectral entropy value is sufficiently small the classifier start detecting the given features samples as seizure condition. Following that, the gradual and sufficient increase in $E_{lowband}$ spectral energy and signal energy features influence the SVM classifier for continuous detection of the seizure features samples correctly. Since the SVM classifier able to identify a large number of seizure features samples starting from 180 seconds, the automated detection of electrical onset was marked at 180 seconds.

Table 4.5: The performance of classifiers using all three significant features on individual seizure recordings (Scheme 1)

Seizure ID	Sensitivity (%)					Specificity (%)				
	LDA	NB	DT	SVM	KNN	LDA	NB	DT	SVM	KNN
P1_s1	52	51	76	79	76	82	81	84	78	75
P1_s2	36	12	70	53	36	100	99	95	94	93
P1_s3	83	84	90	87	87	99	99	32	82	72
P2_s1	44	37	50	63	71	99	99	99	98	89
P2_s2	55	55	73	71	74	99	99	94	97	90
P2_s4	49	48	60	60	58	99	99	99	99	95
P3_s1	45	40	78	71	72	100	99	97	97	95
P4_s1	86	100	97	99	97	69	23	20	42	31
P5_s1	53	55	82	78	78	99	98	83	89	85
P7_s1	53	41	83	73	80	99	97	95	95	94
P7_s2	65	47	87	89	93	100	99	88	91	88
P7_s3	73	62	95	91	90	99	99	94	99	94
P8_s1	93	87	96	88	89	93	91	74	85	78
P8_s3	56	60	73	74	69	99	99	98	98	92
P11_s1	38	35	76	64	65	100	98	97	97	95
P11_s2	34	31	71	60	61	100	99	96	97	95
P11_s3	48	39	81	69	75	98	97	94	95	95
P12_s1	99	100	100	100	100	70	61	27	50	32
P13_s1	45	57	58	59	56	99	98	85	85	83
P14_s1	65	89	69	94	86	84	60	70	55	48
P15_s1	97	78	99	100	100	16	59	01	0	01
P16_s1	73	81	84	83	85	92	85	73	79	69
P17_s1	77	75	96	90	90	98	94	42	80	55
P17_s2	65	58	89	84	85	99	97	84	96	86
P18_s1	41	30	66	66	59	99	98	89	91	93
P19_s1	19	40	46	60	46	91	88	72	71	65
P20_s1	59	62	86	82	75	99	99	98	97	92
P20_s2	67	73	81	87	83	95	90	66	75	70
P21_s1	49	54	78	81	67	73	59	31	35	44

Note: Pn_sm - n^{th} patient m^{th} seizure recording; LDA-Linear Discriminant Analysis; NB-Naive Bayes; DT-Decision Tree; SVM-Support Vector Machine; KNN-K Nearest Neighbor; Sensitivity=TS/(TS+FN); Specificity=TN/(TN+FS); TS-True Seizure; TN-True Normal; FS-False Seizure; FN-False Normal

Similarly for all 29 seizure recordings the detection performance of SVM classifiers to detect the normal and seizure features samples were analyzed. The same method was adapted to determine the detection performance of other classifiers – LDA, NB, DT and KNN to detect the given normal and seizure features samples. Considering all three features of four channels extracted from EEG normalized using seizure specific maximum and minimum values (Scheme 1), the detection performance of classifiers is tabulated in Table 4.5. The classifiers identified the given normal and seizure features samples of each seizure recording with different specificity and sensitivity values respectively. The average performance measures of each classifier across all 29

Table 4.6: The average performance of classifiers trained using all three features (Scheme 1)

Classifier Parameter	LDA	NB	DT	SVM	KNN
Sensitivity(%)	59	58	79	78	76
Specificity(%)	91	89	75	81	76
Accuracy(%)	75	73	76	79	75
Error Rate(%)	25	27	24	21	25

Note: Para - Parameters; Clas - Classifiers; Sensitivity=TS/(TS+FN); Specificity=TN/(TN+FS); Accuracy=(TS+TN)/(Total no. of samples); Error rate=1-accuracy; TS-True Seizure; TN-True Normal; FS-False Seizure; FN-False Normal

seizure recordings is given in Table 4.6. The LDA and NB classifiers performed with sensitivity (specificity) of 59% (91%) and 58% (89%), respectively. The performance of other three (DT, SVM and KNN) classifiers is very similar. The DT and KNN classifiers scored the detection accuracy of 76% and 75%, with sensitivity (specificity) of 79% (75%) and 76% (76%), respectively. The SVM classifier with overall sensitivity (specificity) of 78% (81%) and accuracy of 79%, appeared to be more suitable for the seizure detection system in clinical study.

Among the 29 seizures, four seizures (P4_s1, P12_s1, P15_s1, P21_s1) yielded low specificity of identifying the normal features samples. A typical example in these group, three prominent features extracted from P4_s1 seizure recording, and their corresponding SVM classifier output is shown in Fig. 4.23. The electrical onset of P4_s1 seizure was visually identified by the Neurologist at 159 seconds from left temporal lobe at T1 electrode as shown in Fig. 3.18. The 5 minutes recording of P4_s1 seizure consists of 159 normal features samples and 138 seizure features samples. The SVM classifier identified 99% of the seizure features samples correctly where as only 42% of the normal features samples were identified correctly. Comparing the three features profiles and their corresponding SVM output of P4_s1 seizure shown in Fig. 4.23, the decrease in spectral entropy and increase in $E_{lowband}$ spectral energy due to the non-seizure activity of the brain produced more false seizure detections. Similarly in this group, the normal features samples of P15_s1 seizure, were not identified by LDA, DT, KNN and SVM classifiers. Since the spectral entropy samples of normal EEG overlapped with the features samples of seizure EEG, all the normal features samples were classified as seizure condition, and it was unable to determine its electrical

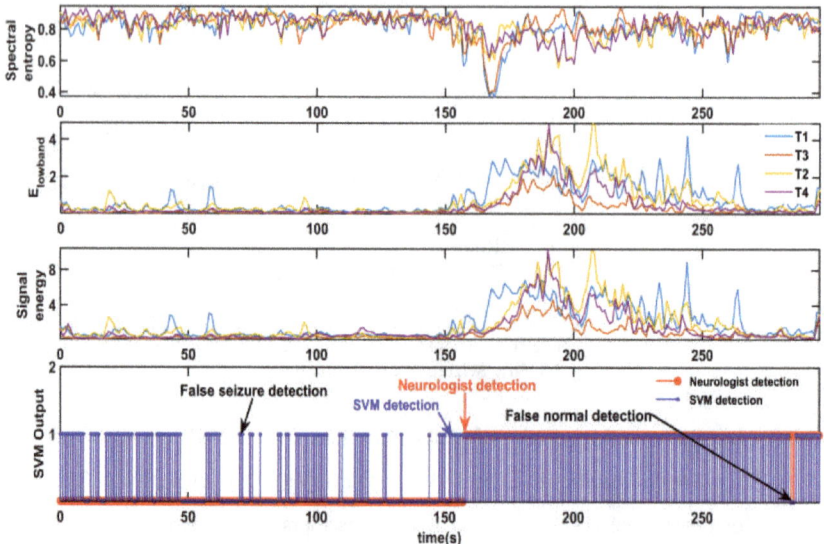

Figure 4.23: The significant features extracted from P4_s1 seizure normalized using channel specific maximum and minimum values and their corresponding SVM classifier output. The seizure electrical onset was visually identified at 159 seconds from left temporal lobe T1 electrode.

onset. Among the five classifiers, NB classifier only detected the electrical onset of P15_s1 seizure at 144 seconds with 59% (78%) of specificity (sensitivity) performance measures.

For all 29 seizures, the automated detection of electrical onset was marked and compared with their visual detection by the Neurologists. The visual and automated detection of electrical onset 29 seizures by all the five classifiers trained using spectral entropy, $E_{lowband}$ spectral energy, and signal energy features are given in Table 4.7. When only considering the seizures detected before their clinical onset, the electrical onset detection latency of each classifier was calculated and reported in Table 4.7. The LDA classifier identified the onset of twenty one seizures with average detection latency of 9 seconds (ranging from -19 seconds to 40 seconds with the median of 7 seconds). The NB classifier identified the onset of eighteen seizures with average latency of 8 seconds (ranging from -19 seconds to 40 seconds with the median of 4.5 seconds). The DT classifier identified the onset of twenty six seizures with average latency of 4.7 seconds (ranging from -20 seconds to 34 seconds with the median of 2 seconds). The KNN classifier identified the onset of twenty six seizures with average latency of

Table 4.7: Detection of electrical onset using all three features (Scheme 1)

Seizure ID	Visual detection (s)		Automated detection (s)								
	EO	CO	LDA	LDA Latency	NB	NB Latency	DT	DT Latency	SVM	SVM Latency	KNN	KNN Latency
P1_s1	138	180	147	9	153	15	141	3	138	0	138	0
P1_s2	143	180	147	4	162	19	145	2	147	4	147	4
P1_s3	165	180	168	3	169	4	165	0	166	1	166	1
P2_s1	174	180	192	-	194	-	194	-	180	-	181	-
P2_s2	159	180	174	15	174	15	174	15	164	5	168	9
P2_s4	142	180	149	7	152	10	152	10	142	0	142	0
P3_s1	152	180	166	14	ND	-	164	12	164	12	168	16
P4_s1	159	180	153	-6	148	-11	149	10	153	6	152	-7
P5_s1	165	180	167	2	167	2	164	-1	164	1	164	-1
P7_s1	175	180	188	-	ND	-	176	1	175	0	176	1
P7_s2	157	180	186	-	186	-	153	-4	156	-1	153	-4
P7_s3	173	180	ND	-	ND	-	182	-	182	-	181	-
P8_s1	177	180	173	-4	173	-4	170	-7	173	-4	173	-4
P8_s3	121	180	161	40	161	40	150	29	150	29	151	30
P11_s1	134	180	149	15	ND	-	136	2	136	2	137	3
P11_s2	114	180	129	15	ND	-	116	2	116	2	117	3
P11_s3	176	180	ND	-	ND	-	178	2	178	2	179	3
P12_s1	162	180	143	19	143	-19	142	-20	142	-20	142	-20
P13_s1	161	180	180	-	180	-	177	16	176	15	176	15
P14_s1	125	180	134	9	125	0	126	1	125	1	126	1
P15_s1	145	180	AS	-	144	-1	AS	-	AS	-	AS	-
P16_s1	163	180	168	5	164	1	164	1	168	5	165	2
P17_s1	163	180	182	-	182	-	163	0	166	3	163	0
P17_s2	138	180	157	19	158	20	141	3	141	3	141	3
P18_s1	141	180	146	5	ND	-	142	1	142	1	144	3
P19_s1	114	180	135	21	130	16	130	16	130	16	134	20
P20_s1	134	180	136	2	139	5	146	12	148	14	149	15
P20_s2	167	180	167	0	168	1	169	2	166	1	169	2
P21_s1	111	180	145	34	144	33	145	32	138	32	140	29
Mean				9		8		4.7		4		5
Median				7		4.5		2		2		2.5
Minimum				-19		-19		-20		-20		-20
Maximum				40		40		34		29		32

Note: Pn_sm - n^{th} patient m^{th} seizure recording; ND - Not Detected - the seizure feature samples only at the end of seizure episode were classified correctly; AS - All the normal features samples were classified as Seizure condition

5 seconds (ranging from -20 seconds to 32 seconds with the median of 2.5 seconds). The SVM classifier identified the onset of twenty six seizures with average latency of 4 seconds (ranging from -20 seconds to 29 seconds with the median of 2 seconds). The two seizures, P2_s1 and P7_s3 were identified by all the classifiers only at their clinical onset. The electrical onset of P15_s1 seizure was only identified by the NB classifier. The other classifiers identified all the normal features samples of P15_s1 as seizure samples and unable to determine its electrical onset.

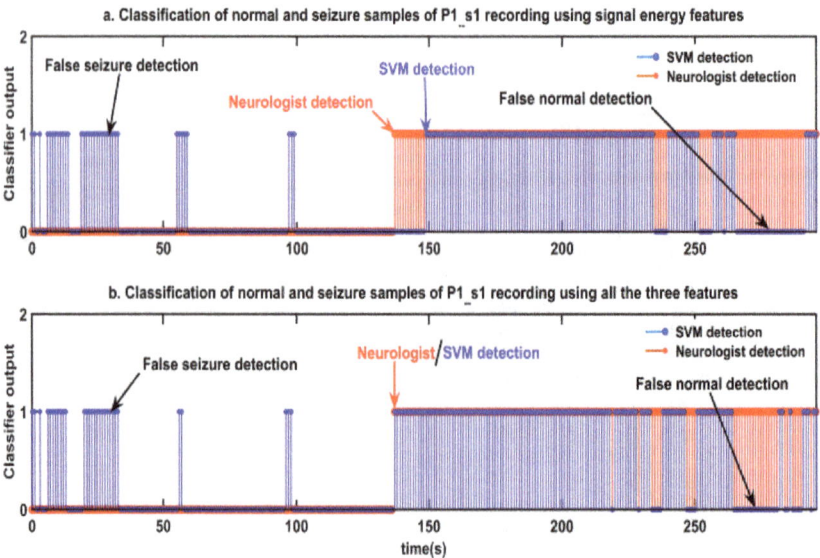

Figure 4.24: Visually identified P1_s1 seizure electrical onset at 138 seconds in right temporal region picked by T2 electrode (Scheme 1). The SVM classifier trained using signal energy features detected the electrical onset at 150 seconds which is shown in (a). The SVM classifier trained using spectral entropy, spectral energy and signal energy detected the electrical onset at 138 seconds as shown in (b).

Incorporating spectral entropy and spectral energy features to train the classifiers improved the performance of electrical seizure onset detection system. The average detection performance of classifiers trained using signal energy features of four channels and, all three features of four channels were presented in Table 4.3 and Table 4.6 respectively. Using all the three features to train the classifier, increased the performance measures meanwhile reduced the electrical seizure onset detection latency. The sensitivity of all the five classifiers trained using all the three features were increased significantly. Particularly, the specificity of DT and KNN classifiers increased significantly to detect the normal features samples correctly. The detection latency of classifiers trained using signal energy features of four channels, and the classifiers trained using all three features of four channels extracted from Scheme 1 normalized EEG were given in Table 4.4 and Table 4.7 respectively.

Considering the detection latency, the mean / median latency of DT, SVM and KNN classifiers decreased significantly to enable the early detection of electrical seizure onset. A typical example of the electrical onset of P1_s1 seizure detected by SVM classifier trained using signal energy, and all three significant features extracted from T1, T3, T2, T4 channels are compared in Fig. 4.24. As shown in Fig. 4.19, the earliest frequency variations in seizure episodes were captured by spectral entropy measures in 3 to 12 Hz which are dominantly influencing the early seizure onset detection. Following the spectral entropy, the $E_{lowband}$ spectral energy sufficiently increased to continuously detect the seizure features samples by the classifier. The signal energy increased at the later part of the seizure episode contributed to correct labeling of the seizure features samples. The SVM classifier trained using three features detected the onset of P1_s1 seizure at 138 seconds which correlates with visual detection (138 seconds) by the Neurologists. The SVM classifier trained using signal energy features detected the electrical onset at 150 seconds which is 12 seconds later the Neurologists' detection. Moreover, SVM classifier trained using signal energy features alone, had the sensitivity and specificity of 66% and 77% respectively. Adding the spectral entropy and spectral energy features to model the classifier increased the sensitivity and specificity values to 79% and 78% respectively. Similar analysis done on all 29 seizure recordings showed that including the spectral entropy and spectral energy features have improved the specificity, sensitivity, and accuracy performance measures of all the five classifiers in detecting the electrical onset of seizure.

The ability of the five classifiers (LDA, NB, DT, KNN and SVM) model to discriminate one class (seizure) from another class (normal) was evaluated using area under the receiver operating characteristics (ROC) curve. The ROC curve is a plot of all of the sensitivity/specificity pairs resulting from continuously varying the decision threshold over the entire range of results observed (Zweig and Campbell, 1993). The ROC plot depicts the overlap between the distribution of two class (seizure and normal) samples by plotting the sensitivity vs (1-specificity) for the complete range of decision thresholds. On the y-axis is sensitivity, or the true-positive rate defined as (number of true-seizure samples)/(number of true-seizure samples + number of false-normal samples). On the x-axis is the false-positive rate, or (1 - specificity) defined as (number of false-seizure samples)/(number of true-normal samples + number of false-seizure results). A classifier model with no overlap in the distribution of two class samples, has an ROC plot that passes through the upper left corner, where the true-positive rate is 1.0, or 100% sensitivity, and the false-positive rate is 0 or 0% specificity. For the classifier model with complete overlap of two distributions has the ROC curve, a diagonal line from the bottom left corner to the top right corner. Qualitatively, closer the plot is to the top left corner, higher the overall accuracy (100% sensitivity and 100% specificity) of the classification test. A guideline to categorize the precision of the test based on area under ROC curve (AUC) value is given as follows: the results are treated to be excellent if the AUC is between 0.9 and 1.0; if the AUC is between 0.8 and 0.89, then the results are regarded to be good; the results are fair for values between 0.7 and 0.79; the results are poor for values between 0.6 and 0.69; if the AUC is between 0.5 and 0.59, then the outcome indicates that the test results are not better than those obtained by chance (Kamath, 2013).

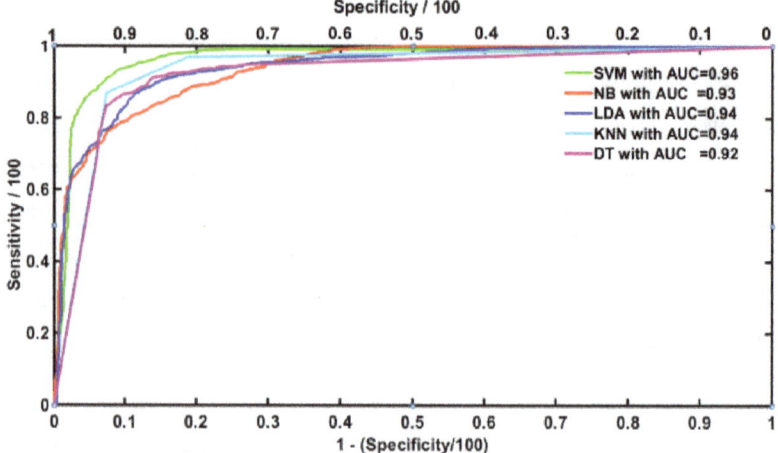

Figure 4.25: The ROC plot of LDA, NB, DT, KNN and SVM classifiers for discriminating seizure and normal EEG condition (Scheme 1).

For ROC analysis, the normal and seizure features samples of all the 29 seizure recordings of 18 patients were grouped and labeled as zeros and ones respectively, and the outliers were removed to test the discriminative ability of LDA, NB, DT, KNN and SVM classifiers' model. Each classifier was trained and tested using 70% and 30% of the data samples respectively. Each classifier was cross-validated to obtain the ROC plot and area under the ROC curve (AUC). The ROC plots of all the five classifiers trained using Scheme 1 normalized EEG for discriminating seizure and normal EEG condition is illustrated in Fig. 4.25. Among the five, SVM classifier scored the highest AUC value of 0.96. The LDA and KNN classifiers scored the AUC value of 0.94. The NB and DT classifiers scored the AUC value of 0.93 and 0.92 respectively. All the five classifiers scored the AUC value close to 1 which ensures the good separability of two class samples distribution for the good sensitivity and specificity performance measures.

4.6 Summary

From recorded EEG, the electrical onset of seizure was detected by using a set of time and frequency domain features extracted from T1, T3, T2 and T4 channels. Among the seven time domain measures, mean, skewness and kurtosis were not showing consistent variations at the electrical seizure onset. Particularly, the approximate entropy and sample entropy features showed the complete overlapping of normal and seizure features samples extracted for the 4 seconds window length EEG. The signal energy showed consistent increase during the seizure episodes in all the 29 recorded seizures. The frequency measures – spectral entropy in 3 to 12 Hz, and spectral energy in 1 to 25 Hz band were included in the feature space. These frequency domain measures showed prominent changes in the distribution of energy at the initial stage of seizure onset. The classifiers trained using spectral entropy, spectral energy in 1 to 25 Hz and signal energy in 1 to 200 Hz band significantly improved the classifiers' performance and detected seizure onset at the earliest.

The focus of this research work is to develop a self-sustaining system, and therefore normalized each recording using seizure specific maximum and minimum values computed from that recording (Normalization Scheme 1). For real time implementation, the computation of maximum and minimum value for upcoming seizure of the patient is unpredictable. Moreover, the selection of seizure specific maximum and minimum value for each patient requires extensive training for classifiers to reconfigure the system for individual patient. A common normalizing factor computed from all 29 seizure recordings is used to normalize each seizure recording. The improved patient independent seizure detection system utilizing the common normalizing factors is presented in the following chapter.

CHAPTER 5

DESIGN OF SEIZURE DETECTION SYSTEM USING NORMALIZATION SCHEME 2

The seizure detection system using EEG Normalization Scheme 1 was discussed in the previous Chapter. In normalization Scheme 1, the selection of maximum and minimum values used for normalizing the EEG is seizure specific of the patient. The selection of these parameters requires the Neurologists input to reconfigure the system for other individuals. As discussed in Chapter 4, the 5 minutes EEG of 29 seizures recorded from 18 patients is considered. In order to enhance the system's independency, the EEG samples of 29 seizure recordings were normalized using common normalizing factors. The suitability of spectral entropy in 3 to 12 Hz band, $E_{lowband}$ spectral energy in 1 to 25 Hz, and signal energy in 1 to 200 Hz features extracted from the EEG normalized using Scheme 2 to detect the electrical seizure onset is demonstrated. The LDA, NB, DT, SVM and KNN classifiers are trained using these features extracted from T1, T3, T2 and T4 channels of the 5 minutes recordings. The performance of the classifiers to detect the given normal features samples and seizure features samples are compared. Moreover, the classifiers performance are tested on cohort data set containing more than one patients' seizure recordings and the detection performance is investigated. The classifier suitable for automated seizure detection system is validated on the entire available length of 29 seizure recordings. In addition to that the suitable classifier is tested on 6 seizures recorded from 4 patients which are not involved in the training phase, and the suitability for clinical routine is analyzed.

5.1 Optimal Features Set

In Normalization Scheme 2, EEG samples of each 5 minutes seizure recording were normalized using the average maximum, and average minimum values. The maximum and minimum values of each channel in the 29 seizure recordings were computed. The average of the 29 maximum values, and average of the 29 minimum values were used to normalize the EEG samples of all 29 seizures as explained in Section. 3.1.6. Since the amplitude of each patient's EEG varied, the signals were normalized using Scheme 2 to make the amplitudes relatively comparable across the patients. As described in Section. 4.2, 4 seconds window sliding 1 second with 3 seconds overlapping of the previous window was used to characterize the EEG. As explained in Section. 3.1.4, each 4 seconds EEG was passed through the high pass filter to obtain the 1 to 200 Hz band limited signal. The notch filters were used to remove the supply line interference and its harmonics at 50 Hz, 100 Hz, 150 Hz and 200 Hz. For each filtered EEG, the most significant features – spectral entropy, $E_{lowband}$ spectral energy and signal energy used for the automated seizure detection system explained in Section. 4.4 were extracted.

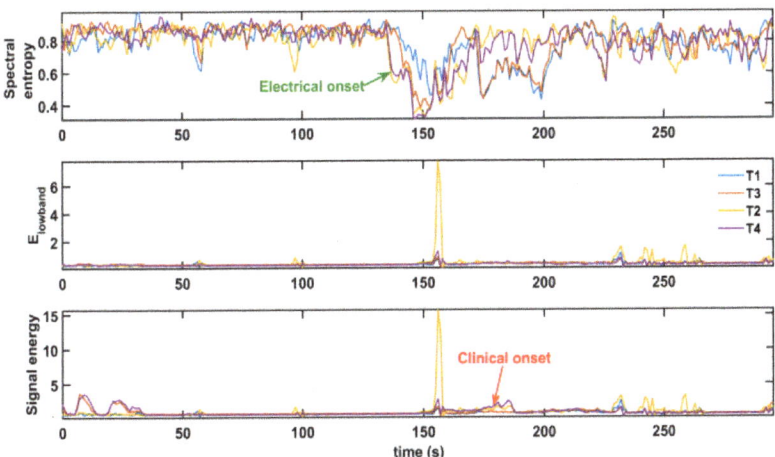

Figure 5.1: The optimal features extracted from P1_s1 seizure recording normalized using Scheme 2.

The profiles of the features extracted from 5 minutes EEG of each seizure recording were analysed. It was observed that the spectral entropy decreases at the electrical onset of seizures. The $E_{lowband}$ spectral energy increases at the initial stage of seizure episode. The signal energy increases during different stages of the seizure episode. The profile of these features correlated with the features extracted from EEG normalized using Scheme 1 as presented in Section. 4.4. The seizure specific maximum and minimum values of 29 seizure recordings and their average maximum and average minimum values were tabulated in Table 3.3. According to these values, the average maximum and average minimum values larger than the seizure specific maximum and minimum values decreased the amplitude of the features extracted from Scheme 2 normalized

EEG. On the other hand, the average maximum and minimum values smaller than the seizure specific maximum and minimum values increased the amplitude of the features extracted from Scheme 2 normalized EEG. The most prominent features of P1_s1 seizure recording normalized using Scheme 2 are shown in Fig. 5.1. The profile of features extracted from EEG normalized using Scheme 2 (Fig. 5.1) correlated with the features extracted from EEG normalized using Scheme 1 (Fig. 4.19), with an amplitude shift according to the normalizing factors. In P1_s1 seizure, the average maximum and average minimum values of T1, T3, and T4 channels are larger than the seizure specific maximum and minimum values of those channels. Hence the amplitude of $E_{lowband}$ and signal energy features decreased in Scheme 2 in the above mentioned channels. Alternatively, the average maximum and average minimum values of T2 channel are smaller than the seizure specific maximum and minimum values. Hence the amplitude of $E_{lowband}$ and signal energy measures increased in normalization Scheme 2 in T2 channel. Likewise the variations in the amplitude of the extracted features of all 29 seizure recordings were analyzed. The impact of three features extracted from the EEG of T1, T3, T2 and T4 channels normalized using Scheme 2 on automated seizure detection were studied.

5.2 Automated Seizure detection

As explained in Chapter 4, the five classifiers LDA, NB, DT, SVM, and KNN were trained for the automated detection of electrical seizure onset. The classifiers were trained using three features – spectral entropy, $E_{lowband}$, and signal energy of the 4 channels T1,T2,T3 and T4 to distinguish the normal and seizure condition. The procedure explained in Section. 4.5 was adopted to prepare the training data to model the classifiers. As illustrated in Fig. 4.20, for each 5 minutes seizure EEG, the features samples were extracted for each 4 seconds window length EEG. The extracted features samples were grouped as normal and seizure samples with respect to the electrical onset visually identified by the Neurologists. The normal and seizure features samples extracted from seizure recordings of N-1 patients were labeled as zeros and ones respectively, to enable the supervised training of classifiers. The features samples extracted from Nth patient's seizure recordings were used for testing the classifiers. As done for Scheme 1 based automated detection system explained in Section. 4.5, the trained classifiers identify the given features sample as normal or seizure condition according to the given 1x12 features set.

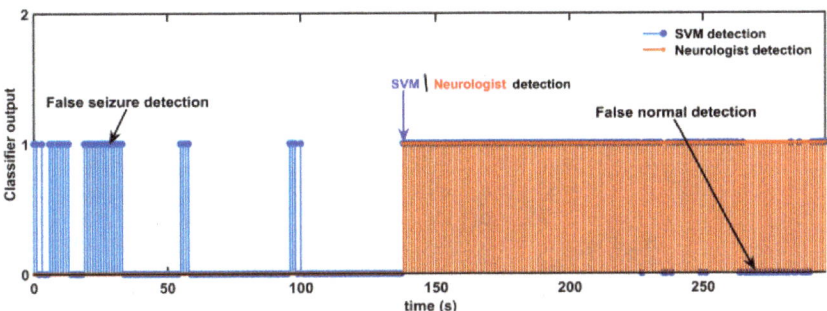

Figure 5.2: The electrical onset detection of seizure P1_s1 by neurologists and SVM trained using three extracted from EEG normalized using Scheme 2. The normal features samples are represented as '0' and the seizure features samples are represented as '1'. Misclassification of the normal and seizure features samples are marked as 'False seizure detection' and 'False normal detection', respectively.

The visual and SVM-based identification of the electrical onset of P1_s1 seizure is shown in Fig. 5.2. The P1_s1 seizure recording consists of 138 normal features samples and 159 seizure feature samples. The SVM classifier identified 75% of the normal features samples as normal and 81% of the seizure features samples as seizure as reported in Table 5.1. The misclassification of normal and seizure features samples are represented as false seizure detections and false normal detections respectively. The SVM classifier output for P1_s1 seizure shown in Fig. 5.2 was compared with the prominent features of P1_s1 extracted from Scheme 2 normalized EEG shown in Fig. 5.1. It was observed that the increase in signal energy and decrease in spectral entropy values during the initial seizure recording (1 to 25 seconds) influence more false seizure detections. Meanwhile, the decrease in spectral entropy at the initial stage of the seizure episode influence the earliest detection of electrical seizure onset by the SVM classifier. Following the decrease in spectral entropy, the increase in $E_{lowband}$ and signal energy values influence the classifier for continuous detection of the seizure cycle. Being able to detect the seizure features samples continuously starting from 138 seconds, the automated detection of electrical onset was marked at 138 seconds. For P1_s1 seizure, the SVM identified seizure onset is coincide with the seizure onset visually identified by the Neurologists as shown in Fig. 5.2.

Table 5.1: Performance of classifiers on individual seizure recordings (Scheme 2)

Seizure ID	Sensitivity (%)					Specificity (%)				
	LDA	NB	DT	SVM	KNN	LDA	NB	DT	SVM	KNN
P1_s1	41	44	71	81	74	93	80	72	75	74
P1_s2	55	15	61	72	61	99	100	96	95	97
P1_s3	43	40	75	77	72	99	100	96	90	98
P2_s1	78	77	89	91	92	99	100	87	94	94
P2_s2	81	78	93	92	88	99	100	92	97	91
P2_s4	70	79	90	90	81	100	100	99	99	98
P3_s1	73	79	100	98	97	99	97	79	90	90
P4_s1	17	2	47	47	47	100	100	99	99	97
P5_s1	34	32	61	68	58	99	99	94	92	93
P7_s1	74	78	94	97	95	96	91	70	85	80
P7_s2	85	64	91	100	87	96	100	89	94	92
P7_s3	81	69	94	93	89	100	99	84	94	87
P8_s1	77	82	96	99	94	94	92	81	89	88
P8_s3	63	72	89	83	83	99	100	88	95	93
P11_s1	70	77	98	98	93	98	94	78	89	92
P11_s2	67	74	96	95	91	99	94	78	89	90
P11_s3	70	78	97	97	93	95	89	73	84	86
P12_s1	51	31	86	94	86	88	88	65	76	73
P13_s1	4	34	42	45	47	99	99	90	89	88
P14_s1	44	16	81	66	67	94	99	66	81	69
P15_s1	100	78	100	100	99	4	67	3	0	3
P16_s1	21	33	80	67	61	97	97	82	87	88
P17_s1	74	28	78	81	72	100	100	93	91	93
P17_s2	59	21	66	78	71	100	100	94	99	99
P18_s1	42	28	67	66	64	98	98	91	90	93
P19_s1	14	14	58	45	46	88	93	81	79	78
P20_s1	17	6	62	60	52	100	100	98	99	100
P20_s2	24	16	56	58	40	99	100	96	99	98
P21_s1	29	17	82	73	70	91	96	32	53	48

Note: Pn_sm - n^{th} patient m^{th} seizure recording; LDA-Linear Discriminant Analysis; NB-Naive Bayes; DT-Decision Tree; SVM-Support Vector Machine; KNN-K Nearest Neighbour; Sensitivity=TS/(TS+FN); Specificity=TN/(TN+FS); TS-True Seizure; TN-True Normal; FS-False Seizure; FN-False Normal

All the five classifiers were tested on 29 seizure recordings using leave one patient out approach. The detection performance of each classifier was evaluated using the performance measures given in Eq. (4.17). The performance measures of each classifier on individual seizure recordings organized from patient independent environment is given in Table 5.1. Among the five classifiers, LDA and NB classifiers scored very high specificity of identifying the given features samples as normal, and low sensitivity of identifying the given features samples as seizure condition. The other three classifiers DT, SVM and KNN scored good sensitivity and specificity performance measures of identifying the seizure and normal features samples respectively. Among the 29 seizure recordings, these three classifiers yielded low sensitivity of identifying the electrical onset of P4_s1, P13_s1, P19_s1 and P20_s2 seizures. In all these four seizures, the spectral entropy showed significant variations at the electrical onset of seizure. The decrease in spectral entropy at the onset of seizure, influence the classifiers to detect the seizure features samples as seizure condition. Contradictorily, in P13_s1, and P20_s2 seizures the significant changes in the signal energy features were observed only at their clinical onset. Moreover the P4_s1, P13_s1, P19_s1 seizures showed no significant variations in the E_{lowband} energy profiles. Due to the lack of significant variations in E_{lowband} and signal energy features during the seizure activity, the detection of these four seizures resulted in more false normal samples in the later part of the seizure episode. Analyzing the specificity measures of all 29 seizures, P15_s1 and P21_s1 seizures were identified with low specificity by DT, SVM and KNN classifiers. Particularly in P15_s1 seizure recording, all the normal samples were classified as seizure samples, it was not able to determine the electrical onset of P15_s1 seizure.

Table 5.2: The average performance of classifiers on independent patient data (Scheme 2)

Parameter \ Classifier	LDA	NB	DT	SVM	KNN
Sensitivity(%)	54	47	79	80	75
Specificity(%)	94	96	81	86	85
Accuracy(%)	75	71	80	83	80
Error Rate(%)	25	29	20	17	20

Note: Para - Parameters; Clas - Classifiers; Sensitivity=TS/(TS+FN); Specificity=TN/(TN+FS); Accuracy=(TS+TN)/(Total no. of samples); Error rate=1-accuracy; TS-True Seizure; TN-True Normal; FS-False Seizure; FN-False Normal.

Good performance of an automated system is ensured by high sensitivity and high specificity performance measures of the classifier. The classifiers identified the given normal and seizures features samples of each seizure recording with different specificity and sensitivity values respectively. The average performance measures of each classifier across all the 29 seizure recordings is given in Table 5.2. The LDA and NB classifiers performed with the sensitivity (specificity) of 54% (94%) and 47% (96%), respectively. The performance of the other three (DT, SVM and KNN) classifiers is very similar. The DT and KNN classifiers scored the accuracy of 80%, with sensitivity (specificity) of 79% (81%) and 75% (85%), respectively. Among the five classifiers, SVM scored high accuracy of 83% with sensitivity and specificity measures of 80% and 86% respectively.

Table 5.3: Detection of electrical onset of seizures using all three features (Scheme 2)

Seizure ID	Visual detection (s)		Automated detection (s)								
	EO	CO	LDA	LDA Latency	NB	NB Latency	DT	DT Latency	SVM	SVM Latency	KNN	KNN Latency
P1_s1	138	180	143	5	153	15	145	7	138	0	139	1
P1_s2	143	180	147	4	163	20	145	2	145	2	147	4
P1_s3	165	180	168	3	182	-	166	1	166	1	168	3
P2_s1	174	180	181	-	189	-	180	-	180	-	180	-
P2_s2	159	180	164	5	170	11	164	5	164	5	166	7
P2_s4	142	180	143	1	147	5	141	-1	141	-1	141	-1
P3_s1	152	180	164	12	164	12	152	0	153	1	155	3
P4_s1	159	180	ND	-	AN	-	166	7	166	7	166	7
P5_s1	165	180	171	6	168	3	164	-1	164	-1	165	0
P7_s1	175	180	176	1	177	2	178	3	177	2	177	2
P7_s2	157	180	163	6	173	16	155	-2	157	0	156	-1
P7_s3	173	180	182	-	214	-	182	-	182	-	182	-
P8_s1	177	180	173	-4	173	-4	172	-5	173	-4	172	-5
P8_s3	121	180	150	29	150	29	150	29	150	29	150	29
P11_s1	134	180	136	2	138	4	127	-7	138	4	138	4
P11_s2	114	180	116	2	118	4	107	-7	118	4	118	4
P11_s3	176	180	178	2	180	-	169	-7	180	-	180	-
P12_s1	162	180	ND	-	ND	-	162	0	162	0	162	0
P13_s1	161	180	AN	-	180	-	179	18	179	18	180	-
P14_s1	125	180	133	8	ND	-	125	0	127	2	127	2
P15_s1	145	180	AS	-	141	-4	AS	-	AS	-	AS	-
P16_s1	163	180	ND	-	185	-	164	1	175	12	168	5
P17_s1	163	180	182	-	237	-	166	3	166	3	166	3
P17_s2	138	180	157	19	210	-	142	4	143	5	143	5
P18_s1	141	180	143	2	217	-	142	1	143	2	143	2
P19_s1	114	180	ND	-	ND	-	127	13	135	21	133	19
P20_s1	134	180	ND	-	AN	-	134	0	136	2	134	0
P20_s2	167	180	188	-	ND	-	167	0	167	0	167	0
P21_s1	111	180	ND	-	ND	-	143	32	143	32	143	32
Mean				6.1		8.7		3.7		5.8		5.2
Median				4		5		1		2		3
Minimum				-4		-4		-7		-4		-5
Maximum				29		29		32		32		32

Note: Pn_sm - n^{th} patient m^{th} seizure recording; ND-Not Detected - the seizure feature samples only at the end of seizure episode were classified correctly; AN- All the patterns were detected as normal; AS-All the patterns were detected as seizure

The classifiers able to identify the given normal and seizure features samples of a group of seizure recordings correctly. The automated detection of electrical onset was marked at the time from which the classifiers continuously detect the given seizure features samples correctly. The automated detection of electrical onset of 29 seizures by all the five classifiers were marked and compared with the visual detection by the Neurologists. The automated and visual identification of electrical onset of all 29 seizure recordings and the latency to detect the electrical seizure onset were tabulated in Table 5.3. The latency is defined as the time difference between the detection of electrical onset by the Neurologists and automated detection system. The negative sign of latency define automated seizure detection happens prior to the visual detection. On the other hand, the positive sign for detection latency define the automated seizure detection happens after the visual detection done by the Neurologists. Considering only the seizures detected before their clinical onset, the following observations are listed. The LDA classifier identified the electrical onset of seventeen seizures with an average latency of 6.1 seconds (ranging from -4 second to 29 seconds with the median of 4 seconds). The NB classifier identified the onset of thirteen seizures with an average latency of 8.7 seconds (ranging from -4 seconds to 29 seconds with the median of 5 seconds). The DT classifier identified the onset of twenty six seizures with an average latency of 3.7 seconds (ranging from -7 second to 32 seconds with the median of 1 seconds). The KNN classifier identified the onset of twenty four seizures with an average latency of 5.2 seconds (ranging from -5 second to 32 seconds with the median of 3 seconds). The SVM classifier identified the electrical onset of twenty five seizures with an average latency of 5.8 seconds (ranging from -4 second to 32 seconds with the median of 2 seconds).

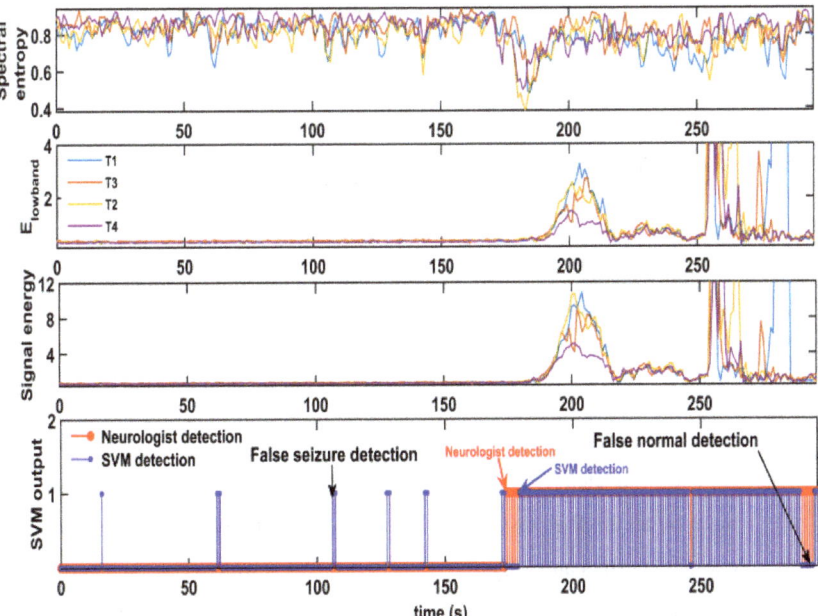

Figure 5.3: The significant features extracted from P2_s1 seizure normalized using Scheme 2 and their corresponding SVM output. The electrical seizure onset visually identified in right temporal region at 174 seconds at T2 electrode.

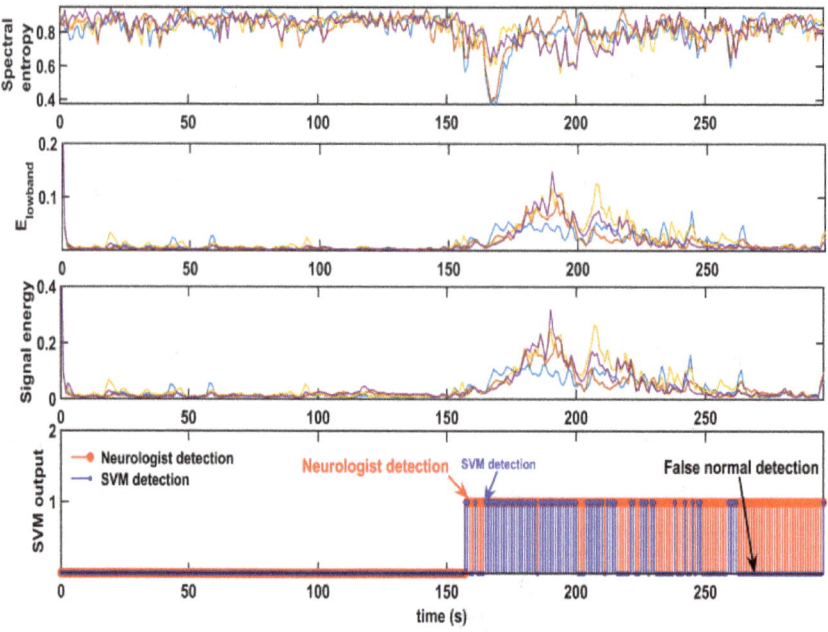

Figure 5.4: The significant features extracted from P4_s1 seizure normalized using Scheme 2 and their SVM detection. The electrical seizure onset visually identified in left temporal region at 159 seconds at T1 electrode.

Among the 29 seizures, electrical onset of a group of seizures were not identified by the classifiers. The LDA classifier was unable to detect the onset of P4_s1, P12_s1, P13_s1, P16_s1, P19_s1, P20_s1 and P21_s1 seizures. Particularly, all the seizure samples of P13_s1 recording were identified as normal samples by the LDA classifier. Similarly, the onset of P12_s1, P14_s1, P19_s1, P20_20 and P21_s1 seizures were not detected by the NB classifier. Moreover all the seizure samples of P4_s1 and P20_s1 were identified as normal samples. In P15_s1 seizure recording, all the normal samples were classified as seizure samples by LDA, DT, SVM and KNN classifiers. Hence it is not able to mark the electrical onset of P15_s1 seizure by the classifiers. Considering the detection latency, P2_s1 and P7_s3 seizures were identified only after their clinical onset by all the five classifiers. The significant features, and their corresponding SVM output of normal and seizure condition of P2_s1 seizure recording is shown in Fig. 5.3. The features were extracted from EEG normalized using the average maximum and minimum values. The average maximum and minimum values are quite smaller than P2_s1 seizure specific maximum and minimum values. Hence the energy level of EEG normalized using Scheme 2 is larger than the energy level of EEG normalized using Scheme 1 as given in Fig. 4.22. The increased signal energy slightly increased the false seizure detection rate, and decreased the false normal detection rate. On the conversely, P4_s1 seizure has the average maximum and average minimum values which are larger than the seizure specific maximum and minimum values. Hence the energy level of EEG normalized using Scheme 2 is lower than the energy level of EEG normalized using Scheme 1 as given in Fig. 5.4 and Fig. 4.23 respectively. The decreased energy level produced 100% specificity of identifying the normal features samples of P4_s1 seizure recording. Comparing the profile of three features of four channels and the classifiers' output, it was observed that the amplitude of $E_{lowband}$ spectral energy and signal energy features has more impact on the false normal and false seizure detection of the given features samples.

5.3 Performance of classifiers on Cohort dataset

The performance of the classifiers was evaluated by another method that used 2/3 of patients (12 patients) for training and the remaining 1/3 of patients (6 patients) was dedicated for testing the classifiers (Kafashan et al., 2017). In Set 1, the seizures recorded from first 12 patients (from P1 to P15) were used for training the classifiers and the seizures recorded from another 6 patients (from P16 to P22) were used for testing the performance of classifiers. Set 2 used the seizures recorded from 12 patients (from P2 to P16) for training and the seizures recorded from other 6 patients (from P17 to P22, and P1) were dedicated for testing the classifiers. Similarly, in a cyclic manner, 12 patients' seizure recordings were used for training and the remaining 6 patients' recordings were used for testing the classifiers. Among the large number of possible combinations, the performance of classifiers were evaluated only on 18 set of patients cohort. The number of seizure recordings used for training and testing the classifiers in each cohort set is reported in Table 5.4. The performance of the classifiers on 18 set of patient cohorts is tabulated in Table 5.4.

The average performance of the classifiers on the 18 sets of cohort data is tabulated in Table 5.5. Comparing the average performance measures of the five classifiers on cohort data sets (Table 5.5) and individual patient data sets (Table 5.2), LDA

Table 5.4: The performance of classifiers on cohort patient recordings

Test set	No. of seizures		Sensitivity (%)					Specificity (%)				
	Training	Test	LDA	NB	DT	SVM	KNN	LDA	NB	DT	SVM	KNN
Set 1	21	8	52	32	66	58	58	90	98	89	90	90
Set 2	19	10	58	35	62	62	59	88	97	91	88	91
Set 3	18	11	54	45	70	67	62	93	97	91	91	93
Set 4	18	11	58	49	73	75	72	93	97	88	89	87
Set 5	18	11	59	49	72	76	69	94	98	90	91	92
Set 6	18	11	63	56	77	77	73	92	97	89	90	91
Set 7	18	11	68	62	78	79	77	93	97	90	91	91
Set 8	18	11	72	71	89	86	85	94	97	82	93	91
Set 9	18	11	71	75	89	86	84	94	92	71	91	87
Set 10	18	11	67	76	90	85	82	93	91	79	90	85
Set 11	18	11	64	80	88	85	84	91	91	77	89	85
Set 12	18	11	68	80	91	86	84	90	89	67	87	83
Set 13	20	9	70	82	93	90	89	77	78	60	73	72
Set 14	21	8	66	78	89	87	85	76	76	65	71	72
Set 15	20	9	70	60	82	82	81	70	82	71	70	72
Set 16	22	7	68	56	77	78	77	72	86	74	72	75
Set 17	22	7	68	53	79	76	77	71	84	72	72	73
Set 18	21	8	61	45	70	67	65	77	88	78	79	79

Table 5.5: Average performance of classifiers on cohort patients data

Classifier / Parameter	LDA	NB	DT	SVM	KNN
Sensitivity(%)	64	60	80	78	76
Specificity(%)	86	91	79	84	84
Accuracy(%)	75	75	79	81	80
Error Rate(%)	25	25	21	19	20

and KNN classifiers secured the accuracy of 75% and 80% respectively. Comparing the sensitivity of LDA classifier on individual patient data, the average sensitivity measure on cohort data increased noticeably. Similarly the KNN classifier's sensitivity (specificity) increased (decreased) by 1% on cohort data sets. The sensitivity measure of NB classifier increased largely to the value of 60% on cohort patients data set. On the other hand, the sensitivity measures of SVM and DT classifiers decreased by 1% and 2% respectively on cohort patients data. Among the five classifiers, SVM classifier scored the lowest error rate of 19% in detecting the normal and seizure features samples correctly.

5.4 Performance of LDA-SVM classifier

The five classifiers were trained using three features extracted from four channels to classify the given set of features samples as normal or seizure condition. Among the five classifiers, SVM secured good sensitivity (specificity) of 80% (86%) and 78% (84%) on individual and cohort patients data set respectively. On the other hand, the LDA classifier tested on cohort patients' data obtained the specificity of 86%. Similarly, the LDA classifier tested on individual patient's data achieved very good specificity of 94% in identifying the given normal features samples (Table 5.2). The SVM classifier showing good sensitivity and specificity was combined with LDA algorithm to improve the performance of seizure detection system. This hybrid classifier named LDA-SVM, used LDA algorithm for the selection of best features (channels) among the four features (channels) from each features set and the SVM was used for determining the class label for the given dimension reduced features samples.

The LDA algorithm reduces the dimension of original feature space while preserving as much of normal and seizure features samples discriminatory information possible. In the dimensionality reduction process, LDA algorithm finds the weight matrix (W), that maximize the distance between the mean of two classes (normal and seizure) and minimize the variance of the features samples of the two classes. For the estimated weight matrix (W), the eigen values and and their corresponding eigen vectors are computed. Then the eigen values are sorted in descending order. The eigen vectors corresponding to the eigen values top the list are defining the weight matrix ($W_{reduced}$) for dimensionality reduced features space. The selection of the number of eigen values determine the reduction in original features set. The detailed explanation of the use of LDA algorithm for dimensionality reduction is presented in Appendix A.

The performance of LDA-SVM classifier was tested on cohort patients' data set explained in Section. 5.3. For each set of cohort patients data, seizure recordings of 2/3 of patients were used for training the classifier and the remaining 1/3 of patients' seizure recordings were used for testing the classifier. For each seizure recording used for training the classifier, three features spectral entropy, $E_{lowband}$ spectral energy and signal energy features from T1, T2, T3 and T4 channels were extracted. The normal and seizure features samples extracted from each seizure recording were grouped and labeled as zeros and ones respectively. In LDA-SVM classifier, the 'n' number of feature samples used for training the SVM were passed through the LDA algorithm for dimensionality reduction as shown in Fig. 5.5. Three LDA modules each using different set of features namely $E_{lowband}$ x 4, signal energy x 4 and spectral entropy x 4 were used for best features (channels) selection. The labeled spectral entropy features of T1, T3, T2 and T4 channels were passed through the first LDA module. The LDA module computes the 4x4 size of weight matrix (W) which maximize the discrimination among the normal and seizure classes. For the 4x4 size of weight matrix (W), 4 eigen values and their corresponding eigen vectors were computed. The four eigen values were sorted in descending order. The eigen vector corresponding to the largest eigen value was chosen to define the weight matrix ($W_{reduced}$) with the size of 4x1 for the dimensionality reduced feature space. The other three smallest eigen values were discarded. The 'n' number of spectral entropy feature samples (nx4) passed through the LDA with 4x1 size of weight matrix $W_{reduced}$ reduced the 4 dimensional features space (nx4) into 1 dimensional feature space (nx1). Similarly, the 'n' number of $E_{lowband}$ (nx4) and signal energy (nx4) features samples passed through the 2nd and 3rd LDA modules respectively, reduced the 4 dimensional features samples (nx4) into 1 dimensional feature samples (nx1).

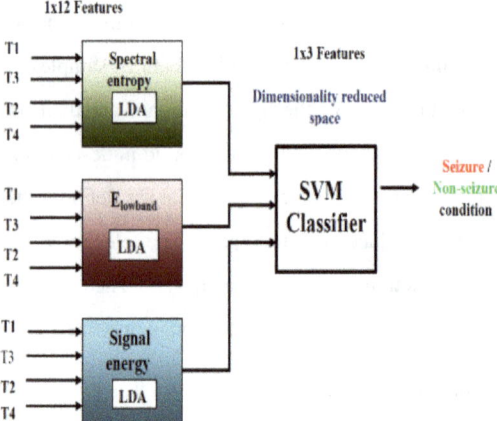

Figure 5.5: LDA based features space dimensionality reduction for SVM classifier on cohort patient data.

With these three LDA classifiers, each using a different set of features, the feature space was reduced from 12 to 3 features (1x$E_{lowband}$, 1x signal energy and 1x spectral entropy) set. The 'n' number of 3-dimensional features samples (nx3) were used to train the SVM classifier. Similarly, features samples extracted from test seizure

Table 5.6: The performance of LDA-SVM classifier on cohort patient recordings

Test set	Sensitivity (%)	Specificity (%)	Accuracy (%)
Set 1	61	91	76
Set 2	66	87	76
Set 3	71	91	81
Set 4	76	90	83
Set 5	77	91	84
Set 6	81	89	85
Set 7	81	92	87
Set 8	90	90	90
Set 9	89	91	90
Set 10	88	90	89
Set 11	89	87	88
Set 12	90	83	86
Set 13	91	72	81
Set 14	88	71	79
Set 15	82	68	75
Set 16	79	73	76
Set 17	76	72	74
Set 18	67	79	73
Average	80	84	82

recordings were passed through these three LDA modules with the same weight matrix $W_{reduced}$ used for the dimension reduction of training features samples, to obtain the 3-dimensional test features samples. The trained SVM classifier determine the class label of the test features samples according to the given 1x3 features samples. The performance of LDA-SVM classifier on a different set of patients cohort as explained in Section. 5.3 is given in Table 5.6. Comparing the average performance measures of SVM and LDA-SVM on 18 cohort data sets, LDA-SVM accuracy increased by 1%, with a 2% rise in the sensitivity and unchanged specificity.

Further, the discriminative ability of SVM and LDA-SVM classifiers' model were analysed using ROC curve. For ROC analysis, the normal and seizure features samples of 18 patients' 29 seizure recordings normalized using scheme 2 were grouped and labeled as zeros and ones respectively, and the outliers were removed. Each classifier was trained and tested using 70% and 30% of the data samples respectively and cross-validated to obtain the ROC plot. The ROC plots of SVM and LDA-SVM classifiers trained using Scheme 2 normalized EEG for discriminating seizure and

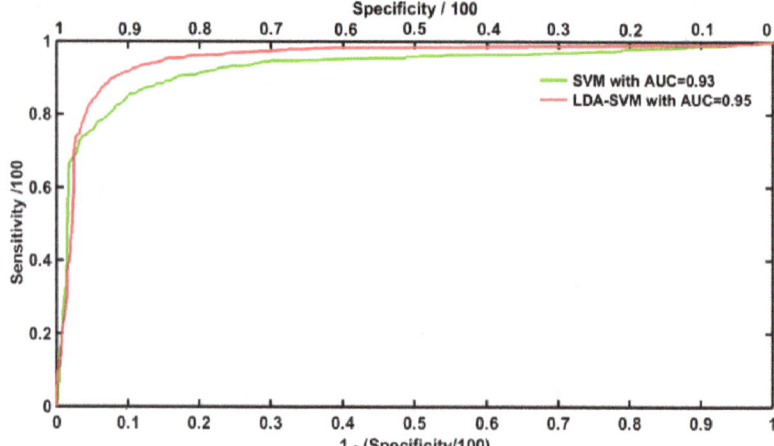

Figure 5.6: The ROC plot of SVM and LDA-SVM classifiers for discriminating seizure and normal EEG condition (Scheme 2).

normal EEG condition is illustrated in Fig. 5.6. The SVM and LDA-SVM classifiers scored the AUC value of 0.93 and 0.95 respectively. Both the classifiers scored the AUC values close to 1 which ensures the good separability of normal and seizure samples distributions.

5.5 Long EEG Recordings

Five classifiers – LDA, NB, DT, SVM and KNN were trained for the automated detection of electrical onset of temporal lobe epileptic seizure. The performance of classifiers tested on individual and cohort patients' data set using patient independent approach was explained in Section. 5.2 and Section. 5.3 respectively. Among the five classifiers, SVM secured good sensitivity and specificity performance measures of identifying the normal and seizure feature samples respectively. Further the LDA-SVM classifier trained using significant features (channels) selected by LDA algorithm showed improved sensitivity performance on cohort patients' data set (Table 5.6). For further analysis the SVM and LDA-SVM classifiers were considered to test the suitability of the proposed system for clinical routine.

While training time, only 5 minutes EEG of 29 seizures recorded from 18 patients (P1 to P22) were used to model the classifiers. On the other hand, the entire length

Table 5.7: Performance of SVM classifier on long EEG recordings

Seizure ID	Sensitivity(%)		Specificity(%)		Detection latency (s)	
	SVM	LDA-SVM	SVM	LDA-SVM	SVM	LDA-SVM
P1_s1	81	83	75	73	0	0
P1_s2	72	75	98	99	2	4
P1_s3	77	80	90	93	1	3
P2_s1	91	93	97	99	-	-
P2_s2	92	93	96	98	5	5
P2_s4	90	91	80	79	-1	-1
P3_s1	98	97	94	95	1	1
P4_s1	47	48	98	98	7	7
P5_s1	68	69	89	89	-1	-1
P7_s1	97	98	86	85	2	0
P7_s2	100	100	96	94	0	-2
P7_s3	93	91	94	96	-	-
P8_s1	99	96	77	74	-4	-4
P8_s3	83	83	76	74	29	29
P11_s1	98	97	97	97	4	2
P11_s2	95	93	97	98	4	2
P11_s3	97	96	96	96	-	2
P12_s1	94	95	78	77	0	-1
P13_s1	45	39	86	87	18	-
P14_s1	66	84	73	68	2	1
P15_s1	100	100	03	02	-	-
P16_s1	67	66	53	70	12	14
P17_s1	81	86	93	93	3	3
P17_s2	78	79	98	98	5	4
P18_s1	66	64	88	90	2	1
P19_s1	45	45	76	77	21	21
P20_s1	60	59	95	71	2	0
P20_s2	58	62	100	99	0	0
P21_s1	70	74	86	87	32	32
P3_Awake	0	0	87	84	-	-
P3_Sleep	0	0	98	97	-	-
P21_Awake	0	0	93	96	-	-
Average Performance	**80**	**81**	**86**	**85**		
Mean					5.8	4.9
Median					2	2
Minimum					**-4**	**-4**
Maximum					**32**	**32**

of available EEG of 29 seizure recordings were used for validating the classifiers trained using the features samples extracted from 5 minutes EEG. The performance of the proposed SVM and LDA-SVM based seizure detection system was validated on entire length of EEG recordings consists of normal (interictal) awake and sleep states preceding the electrical onset of 29 seizures. This analysis was used to test the classifiers' performance to detect the normal features samples extracted from interictal EEG long before the electrical onset of seizure. For all 29 seizure recordings, the normal EEG preceding the electrical seizure onset, with average length of 1167 seconds (ranging from 190 seconds to 2547 seconds with the median of 1262 seconds) were used for examining the specificity performance measures of the classifiers. Apart from the collected 29 seizure recordings, the algorithm was validated with additional three normal (interictal) EEG with awake (P3_Awake, P21_Awake) and sleep (P3_Sleep) episodes recorded from P3 and P21 patients.

For each seizure recording, the EEG of four electrodes T1, T3, T2 and T4 were divided into 4 seconds windows with 3 seconds overlapping with previous window. Each 4 seconds EEG was normalized using Normalization Scheme 2. Each 4 seconds long EEG was passed through the high pass filter to obtain the 1 to 200 Hz band limited signal. The supply line interferences were removed using notch filters at 50 Hz, 100 Hz, 150 Hz and 200 Hz frequencies. As explained in Section. 5.1, the optimal features – spectral entropy, spectral energy and signal energy were extracted for each 4 seconds window length of filtered EEG. The features samples of 29 seizure recordings with additional three normal EEG recordings (P3_Awake, P3_Sleep and P21_Awake) were tested on patient independent SVM classifiers explained in Section. 5.2. The P3_Awake and P3_Sleep recordings were tested on the SVM classifier along with P3_s1 seizure recording. Similarly, the P21_Awake recording was tested on the patient independent SVM classifier along with P21_s1 seizure recording. The specificity performance measure of the SVM classifier tested on normal (interictal) EEG of 29 seizure recordings is tabulated in Table 5.7.

The SVM classifier utilizing the original 12 features set showed more false seizure detection in P16_s1 seizure recording. A 50 seconds EEG segment preceding the onset of P16_s1 seizure is illustrated in Fig. 5.7. The profile of three features influencing the automated detection of the given samples is illustrated in Fig. 5.8. Comparing the P16_s1 EEG presented in Fig. 5.7 and SVM output corresponding the features profiles illustrated in Fig. 5.8, the presence of rhythmic and large amplitude non-seizure patterns (highlighted in magenta color) produced more false seizure samples detection. Apart from that, as explained in previous Section. 5.2, all the normal samples of P15_s1 seizure recording were classified as false seizure samples. All other recordings, scored good specificity in detecting the normal feature samples. Comparing the awake (P3_Awake) and sleep (P3_Sleep) EEG of P3 patient, the SVM classified the feature samples of sleep EEG with 98% of specificity than the feature samples of awake EEG

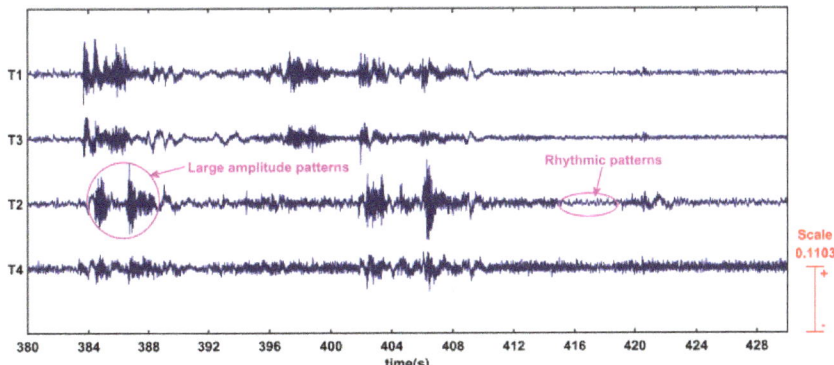

Figure 5.7: A 50 seconds EEG segment preceding the P16_s1 seizure onset.

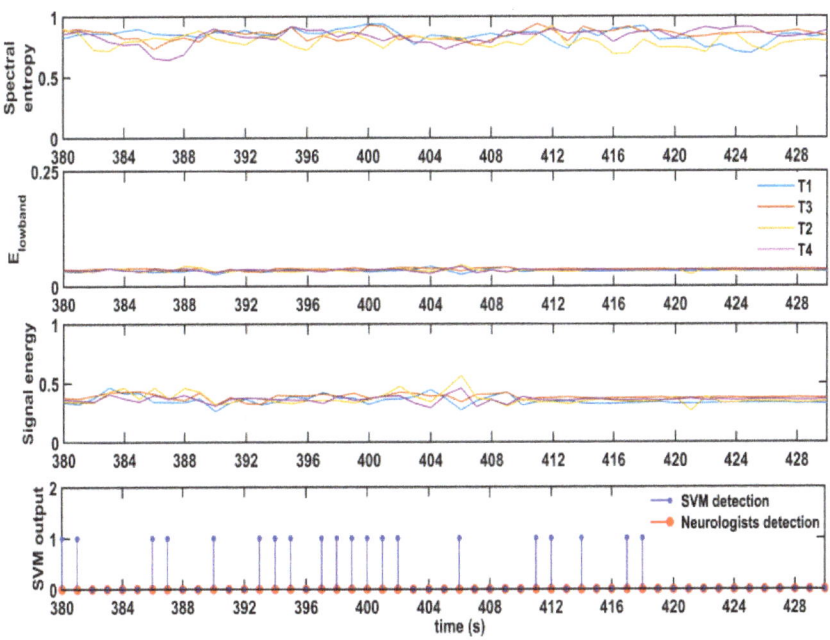

Figure 5.8: The spectral entropy, $E_{lowband}$ and signal energy profile of 50 seconds EEG preceding the P16_s1 onset and their corresponding SVM output.

with 87% of specificity. Similar to the artifacts (the electrical activity recorded from non cerebral site) highlighted in Fig. 5.7, the scalp recorded EEG is highly contaminated by two types of artifacts arise from physiological/biological and non-physiological sources. The physiological artifacts are generated by non cerebral area such as body or head of the patient. The non-physiological artifacts are generated by the equipments used for recording the cerebral activities. The scalp recorded EEG is more prone to physiologic artifacts such as body and head movements, eye movement, electromyogenic (EMG) and sweat artifacts, and non-physiologic artifacts such as electrode movement, electrode cable movement, too much electrode conductive paste, supply line interferences and so on (Zhang *et al.*, 2015). These artifacts confine to a single electrode or spreading over the four electrodes selected by this algorithm play a significant role in determining the given features EEG condition.

Hence, one channel showing good separation among the normal features samples and seizure features samples was selected for automated detection. As explained in Section. 5.2, the normal and seizure features samples of N-1 patients' seizure recordings, were grouped and labeled as zeros and ones respectively. As explained in Section. 5.4, the labeled features samples – spectral entropy, $E_{lowband}$ spectral energy and signal energy of T1, T3, T2 and T4 channels were passed through three LDA modules. Each LDA module select the best feature (channel) among the 4 features (channels) from each group of features set. As explained in Section. 5.4, for each group of features set (spectral entropy, $E_{lowband}$ spectral energy and signal energy) the weight matrix ($W_{reduced}$) with eigen vector corresponding to the largest eigen value was obtained for dimension reduced feature space. Each group of 'n' features samples collected from N-1 patients' recordings were passed through their corresponding weight matrix $W_{reduced}$ to obtain the reduced features set (nx3) to train the classifiers. Similarly, features samples extracted from Nth patient's seizure recordings were passed through these three LDA modules with the same weight matrix $W_{reduced}$ used for the dimension reduction of training features samples, to obtain the 3-dimensional test features samples. The trained SVM classifier determine the class label of the test features samples according to the given 1x3 features samples. Thus the specificity performance of LDA-SVM classifier was tested on individual seizure recordings using patient independent approach and the performance measures are tabulated in Table 5.7. Comparing the performance of SVM and LDA-SVM classifiers, the LDA-SVM showed good specificity of 70% in identifying the normal EEG samples preceding the onset of P16_s1 seizure. On the other hand, the LDA-SVM showed low specificity value of 68% of identifying the normal samples preceding the onset of P14_s1 seizure. For the normal (interictal) EEG of P15_s1 recording, both the classifiers almost identified all the normal features samples as seizure condition. On all other recordings, both the classifiers equally performed good in identifying the features samples of normal EEG.

The sensitivity performance measures of SVM classifier designed using Scheme 2 normalized EEG presented in Table 5.1 is reproduced in Table 5.7. As explained in the above paragraph, the trained LDA-SVM classifier was tested on same length of seizure EEG data explained in Chapter 5. The sensitivity performance measures of LDA-SVM classifier is presented in Table 5.7. The SVM classifier trained using all three features extracted from four channels, scored the average specificity measure of 86% across the 29 seizure recordings with additional 3 normal EEG recording from P3 and P21 patients. The LDA-SVM classifier trained using dimension reduced features samples, scored the average specificity measure of 85% across the 29 seizure recordings with additional 3 normal EEG recordings. Overall, the SVM and LDA-SVM classifiers scored the average accuracy of 83% with the average sensitivity (specificity) of 80%(86%) and 81%(85%) respectively of identifying the given features samples of all 29 seizure recordings. The electrical seizure onset detection latency of each seizure by SVM and LDA-SVM classifiers are compared in Table 5.7. Considering only the seizures detected before their clinical onset, the SVM classifier identified 25/29 seizures with median / mean latency of 2 seconds / 5.8 seconds (ranging from -4 seconds to +32 seconds). The similar analysis performed on LDA-SVM classifier, the electrical onset of 25/29 seizures were identified with median / mean latency of 2 seconds / 4.9 seconds (ranging from -4 seconds to +32 seconds).

5.6 Testing the proposed algorithm on new recordings

The previous section explained the performance of SVM and LDA-SVM classifiers tested on the entire length of available EEG of 29 seizures recorded from 18 patients involved in the development of seizure detection system. Further the proposed patient independent system was tested by using new seizure recordings which are not involved in the classifier model development. The 6 seizures recorded from 4 patients (P22, P23, P24 and P25) described in Section. 3.1 were used for testing the proposed system's performance on unknown seizure recordings. The electrical onset and clinical onset of these six seizures were visually identified by three Neurologists. The electrical onset of seizures identified by the Neurologists were compared with their automated detection by the classifiers, and the detection performance is given in Table 5.8.

As explained in Section. 3.1.3, 6 seizures recorded at the sampling rate of 256 Hz were up-sampled to 400 Hz to ensure the uniform window length across the seizures. Each recording was segmented, normalized and filtered to obtain 1 to 200 Hz band limited signal. The spectral entropy in 3 to 12 Hz, $E_{lowband}$ spectral energy in 1 to 25 Hz, and signal energy in 1 to 200 Hz band – features were extracted for each 4 seconds window length EEG.

As explained in Section. 5.2, the normal and seizure features samples extracted from the 5 minutes EEG of 29 seizures recorded from 18 patients (P1 to P21) were used to train the classifiers. The 12 features extracted from 6 new seizure recordings passed through the trained SVM and LDA-SVM classifiers for automated detection of the given features samples according to the given 1x12 features set. The performance measures of SVM and LDA-SVM for the 6 new recordings were computed and tabulated in Table 5.8. For each seizure recording, the visually identified seizure onset was compared with the onset detected by automated system using SVM and LDA-SVM classifiers.

Table 5.8: The proposed algorithm on new seizure recordings

Seizure ID	$Onset_V$	SVM				LDA-SVM			
		Sensi.(%)	Spec.(%)	$Onset_A(s)$	DL(s)	Sensi.(%)	Speci.(%)	$Onset_A(s)$	DL(s)
P22_s1	1803	27	97	1805	2	29	98	1804	1
P22_s2	1997	33	86	2021	24	18	95	ND	-
P23_s1	45	31	98	51	6	33	100	51	6
P23_s2	164	58	92	172	8	65	91	170	6
P24_s1	205	84	91	207	2	81	90	205	0
P25_s1	341	88	92	340	-1	86	94	341	1

Note: Sensi.- Sensitivity; Spec. - Specificity; $Onset_V$ - Electrical seizure onset visually identified by the Neurologists; $Onset_A$ - Electrical seizure onset detected by the proposed algorithm; ND - Not Detected; DL - Electrical onset detection latency is the time difference between the visual and automated detection of electrical onset of seizure

Comparing the sensitivity and specificity measures of SVM and LDA-SVM classifiers, both of them scored very good specificity in correctly identifying the normal samples preceding the onset of 6 seizures. Among the six new seizures, in the three seizures (P23_s2, P24_s1 and P25_s1), both the classifiers showed large sensitivity values in correctly identifying the seizure features samples. Fig. 5.9 shows the onset of P24_s1 seizure in the right temporal lobe, for four channels of the scalp EEG. The electrical manifestation of seizure started at 205 seconds in T2, the right temporal region with 4 Hz theta band oscillations, and picked-up by T4 electrode in ipsi-lateral temporal region. The P24_s1 seizure recording consists of 204 seconds of normal

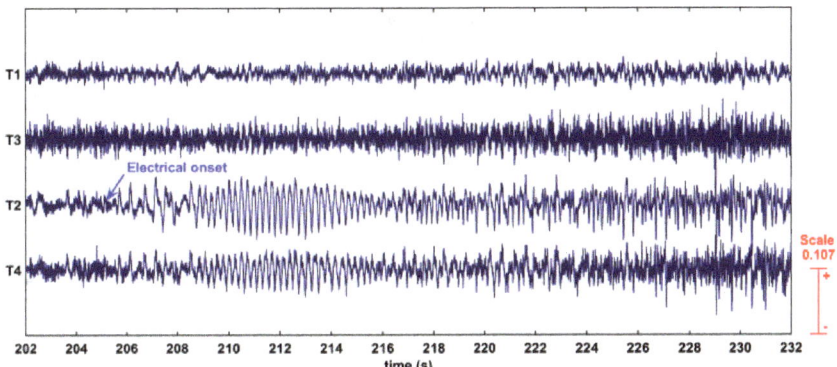

Figure 5.9: Scalp-recorded EEG of the selected electrodes in P24_s1 seizure recording. The seizure onset was visually identified in the right temporal region, and rhythmic theta activity commenced at the T2 electrode (arrow) at 205 seconds.

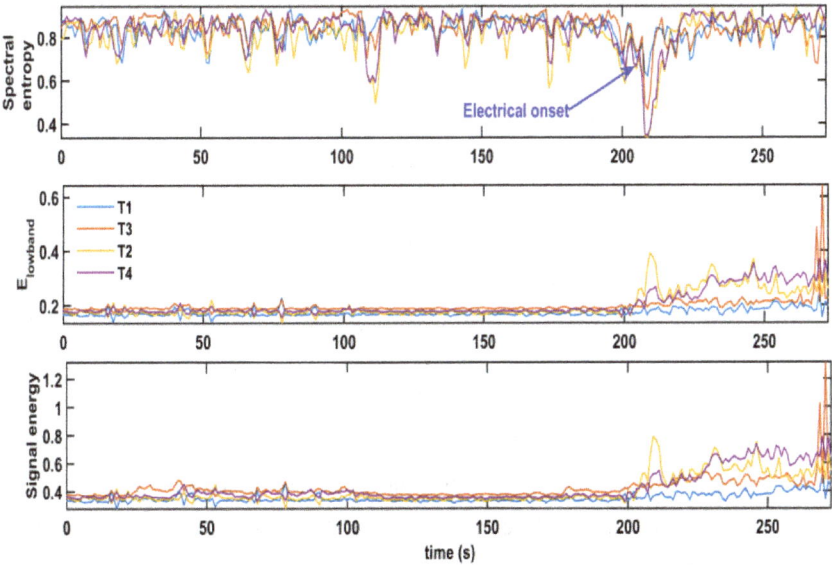

Figure 5.10: The three prominent features extracted from P24_s1 seizure normalized using Scheme 2 maximum and minimum values.

EEG preceding the electrical onset and 70 seconds (length of seizure episode visually identified by the Neurologist) of seizure EEG following the electrical onset. The profile of three prominent features extracted from P24_s1 seizure normalized using Scheme 2 maximum and minimum values is illustrated in Fig. 5.10. The corresponding EEG features samples identification by SVM and LDA-SVM classifiers are shown in

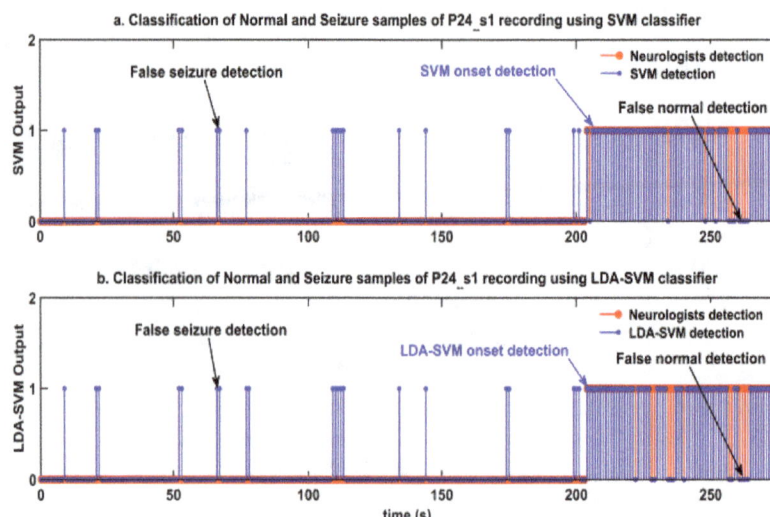

Figure 5.11: The automated identification of normal seizure samples of P24_s1 by SVM and LDA-SVM classifiers.

Fig. 5.11. The electrical onset of P24_s1 seizure detected by SVM and LDA-SVM classifiers is compared. The SVM detected the P24_s1 seizure onset at 207 seconds which is 2 seconds later the visual onset detection (205 seconds) by the Neurologists is shown in Fig. 5.11(a). The LDA-SVM classifier detected the onset of P24_s1 at 205 seconds which correlates with Neurologists detection is shown in Fig. 5.11(b).

Comparing the performance of SVM and LDA-SVM classifiers tested on all 35 seizure recordings, both of them produced good specificity of identifying the normal features samples. Among the 35 seizure recordings, both the classifiers had low sensitivity by continuously identifying the seizure features samples only at the initial stage of P22_s1, P22_s2 and P23_s1 seizure episodes. Comparing three features profiles and their corresponding SVM(LDA-SVM) output of P22_s1 seizure, the classifiers identified the electrical onset with 27%(29%) sensitivity, due to the decrease in spectral entropy at the electrical onset visually identified by the neurologists. The $E_{lowband}$ spectral energy and signal energy features are not showing sufficient increase at the electrical onset of P22_s1 seizure. Similarly for P23_s1 seizure, the decrease in spectral entropy at the beginning of seizure episode contributed most to identify its electrical onset with 31%(33%) sensitivity by SVM(LDA-SVM) classifiers. Alternately, in P22_s2 seizure recording, the significant increase in signal energy at the later part of episode contributed most to identify the seizure samples with sensitivity of 33%(18%) by SVM(LDA-SVM) classifiers.

Considering the long EEG recordings, the False Positive Rate (FPR) and False Negative Rates (FNR) of all 35 seizure recordings were evaluated. The SVM classifier obtained the FPR and FNR of 0.14 and 0.25 in misidentifying the normal and seizure feature samples as seizure and normal respectively. The LDA-SVM classifier obtained the FPR and FNR of 0.14 and 0.24 respectively. Comparing the detection latency of electrical seizure onset identified by these classifiers, the SVM classifier identified 31/35 seizures with median / mean latency of 2 seconds / 6.0 seconds (ranging from -4 seconds to +32 seconds). According to the detection latency of SVM classifier given in Table 5.3, the onset of three seizures – P2_s1, P7_s3 and P11_s3 were identified only after the clinical onset visually identified by the Neurologists. In this case, among the 31 seizures detected by SVM classifier, the electrical onset of P13_s1 seizure detected at 18 seconds latency increased the mean detection latency of SVM classifier to 6.0 seconds. The LDA-SVM classifier identified the electrical onset of 30/35 seizures with median / mean latency of 1.5 seconds / 4.5 seconds (ranging from -4 seconds to +32 seconds). Among the 35 seizures, the onset of three seizures – P2_s1, P7_s3 and P13_s1 were identified only after their clinical onset. The LDA-SVM classifier tested on P22_s2 seizure recording, misclassified all the seizure features samples as normal EEG. Hence, the LDA-SVM classifier was unable to identify the onset of P22_s2 seizure. Even though the LDA-SVM classifier missed the P22_s2 seizure, considering the median / mean detection latency it is recommended for the clinical utility of proposed seizure detection system.

5.7 External Validity

The reproducibility and external validity of the proposed algorithm was investigated on two datasets available from two different repositories. The first dataset is available from a repository created using scalp EEG recordings of ten epilepsy patients collected at Neurology and Sleep Centre (NSC), Hauz Khas, New Delhi. In this dataset, the EEG was recorded using Comet AS40 EEG machine at the sampling rate of 200 Hz. The EEG signals were filtered between 0.5 to 70 Hz and then segmented into pre-ictal, interictal and ictal stages. Each downloadable folder contains fifty MAT-files of EEG time series signals. Each MAT-file consists of 1024 samples of single channel EEG time series data lasting for the duration of 5.12 seconds (Swami *et al.*, 2019). Another dataset is publically available Bonn EEG database (Andrzejak *et al.*, 2001). The Bonn dataset consists of EEG data of five healthy subjects (Set A and Set B) and five epilepsy patients (Set C, Set D and Set E). Set A and B contain seizure-free scalp EEG data recorded from five healthy subjects. During the acquisition, the healthy subjects were relaxed and awake with eyes open (dataset-A) and eyes close (dataset-B). Sets C, D, and E consist of intracranial EEG recorded from epilepsy patients during the pre-surgical diagnosis. The datasets C and D were recorded during seizure-free from electrodes placed within and opposite to the epileptogenic regions, respectively. The dataset E consists of epileptic seizure EEG collected from electrodes placed within the epileptogenic region. Each subset (A, B, C, D, E) of the Bonn database consists of 100 single channel EEG segments, each lasting 23.6 seconds with 4097 samples. Each EEG signal was recorded at the sampling rate of 173.610 Hz.

For the dataset collected from NSC, New Delhi, 50 .MAT files of each folder (pre-ictal, interictal and ictal) was combined into a single time series data. Hence each stage EEG consists of 256 seconds data with 51200 samples. The EEG signals recorded at the rate of 200 Hz were upsampled to 400 Hz to produce the compatibility of the extracted features. The upsampled EEG was passed through the preprocessing module. The 4 seconds window sliding over the 256 seconds long EEG produced 253 features samples for each EEG stage. The extracted features were passed through the SVM and LDA-SVM classifier for the automated classification of the given pre-ictal, interictal and ictal samples. The proposed algorithm consider the 3 features (spectral entropy, $E_{lowband}$ spectral energy, and signal energy) extracted from 4 channels EEG (3x4=12 features) for automated seizure detection. According to the available single channel time series data, the test feature set size is 3x1. Hence to adopt the size of test feature set, for each feature the training feature set was reduced by choosing best channel among the four channel. The ReliefF, a filter based feature selection algorithm was used to choose a best channel discriminating the normal and seizure samples. The training set contributed by the best features selected using ReliefF algorithm was used to train the SVM classifier. Similarly, the LDA-SVM classifier was trained using the same best features selected by using ReliefF algorithm.

Table 5.9: Performance of SVM and LDA-SVM classifiers on Delhi dataset

EEG state	Classification Accuracy (%)	
	SVM	LDA-SVM
Interictal	100	100
Pre-ictal	91	91
Ictal	98	98

Table 5.10: Performance of SVM and LDA-SVM classifiers on Bonn dataset

EEG state	Classification Accuracy (%)	
	SVM	LDA-SVM
Set A (Awake eyes open - Healthy)	99.7	99.7
Set B (Awake eyes closure - Healthy)	66	70
Set C (Within epileptic zone Interictal phase)	78	82
Set D (Opposite to epileptic zone Interictal phase)	68	71
Set E (Ictal iEEG)	100	100

The performance of SVM and LDA-SVM classifier on Delhi dataset is tabulated in Table 5.9. The classification accuracy of pre-ictal, interictal and ictal EEG feature samples are given in Table 5.9. According to the proposed algorithm, the pre-ictal and interictal feature samples are considered as normal and ictal features samples are considered as seizure condition. The SVM and LDA-SVM classifiers identified all the interictal feature samples correctly as normal samples. The pre-ictal and ictal fetaures samples were classified with 91% and 98% accuracy respectively.

Similarly the performance of the algorithm was validated on Bonn dataset consists of scalp and intracranial EEG recorded from normal and epileptic patients. The performance of SVM and LDA-SVM classifiers investigated on Bonn dataset is reported in Table 5.10. The classification accuracy of Set A and Set E EEG feature samples are high as 99.7% and 100% respectively. The SVM and LDA-SVM classifiers scored 78% and 82% accuracy on correctly identifying the feature samples extracted from interictal EEG recorded from within epileptic zone electrodes. Alternately, the SVM and LDA-SVM classifiers scored 68% and 71% accuracy on correctly identifying the feature samples extracted from interictal EEG recorded from electrodes placed on opposite to epileptic zone. Among the five stage EEG data, the SVM and LDA-SVM classifiers showed slightly low value of classification accuracy of 66% and 70% respectively in identifying the awake state with eyes closure.

5.8 Utility in the Clinical study

The proposed patient independent seizure detection system may be utilized in clinical study to assist the Neurologists and Neurotechnologists for automated partition of normal EEG and seizure episodes. This system would be useful in reviewing the long EEG recording, and reduce the review time spent. As described in Section. 3.3 only four channels T1, T3, and T2, T4 were selected to detect the electrical onset of seizures from left and right temporal brain region. The EEG of 4 seconds window with 75% overlapping of previous window ensure the smooth profile of signal characteristics to design the automated system. The schematic of the proposed patient independent seizure detection system is illustrated in Fig. 5.12. The automated seizure detection system acquires the 4 seconds EEG from T1, T3, T2 and T4 channels. The 4 seconds EEG is normalized using average maximum and average minimum values computed from 29 seizure recordings of 18 patients (P1 to P21). The normalized EEG is passed through the high pass filter to obtain the 1 to 200 Hz band limited signal. The notch filters are used to remove the supply line interference and its harmonics. For each 4 seconds windowed EEG, spectral entropy, $E_{lowband}$ spectral energy and signal energy features are extracted. The function of LDA-SVM classifier as explained in Section. 5.4, the 12 features are passed through the LDA-SVM classifier trained using 18 patients'data, for the automated classification of normal and seizure condition. Thus the proposed patient independent system illustrated in Fig. 5.12 could be used in the clinical environment for the automated classification of 4 seconds windowed EEG according to the given 12 features samples.

Continuously getting the features samples of 4 seconds EEG as an example illustrated in Fig. 4.20, for every 1 second the automated seizure detection system label the features samples as normal or seizure EEG. The proposed system able to detect the electrical onset of temporal lobe epileptic seizures at the earliest by utilizing the changes in the spectral entropy and spectral energy features. The increase in signal energy features significantly contribute to detect the given features samples continuously for the entire seizure episode. The earliest electrical onset detection may be helpful to design a therapeutic intervention system to inject Anti Epileptic Drugs (AEDs) to suppress the clinical seizure onset.

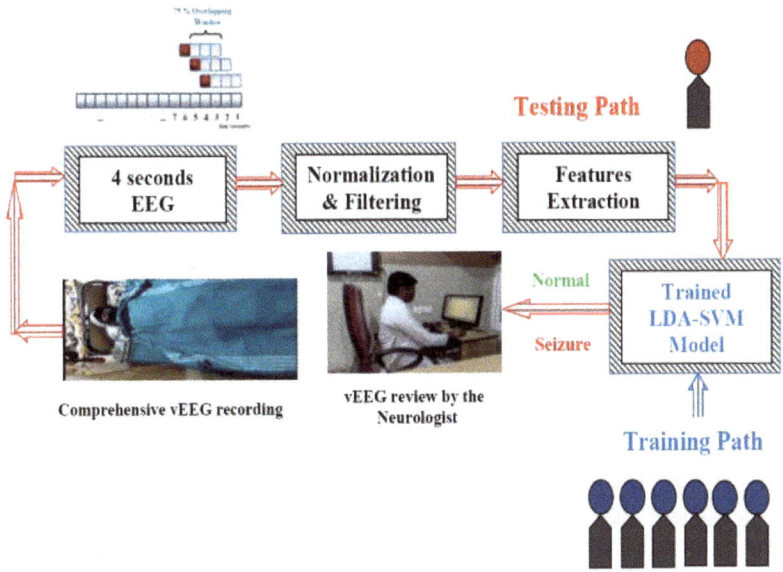

Figure 5.12: The Clinical utility of the proposed seizure detection system.

5.9 Summary

A patient independent seizure detection system utilizing the features extracted from EEG normalized using Scheme 2 is found more suitable for clinical study. The three features extracted from the EEG of four channels (12 features) were used for training the LDA, NB, DT, SVM and KNN classifiers. Among the five classifiers tested on patient independent seizure recordings, SVM classifier had good sensitivity and specificity of 80% and 86% respectively in detecting the normal and seizure EEG. Furthermore, the performance of classifiers tested on 18 sets of cohort patients' data had the sensitivity and specificity of 78% and 84% respectively. Further improving the SVM performance, the features space was reduced using three LDA modules. With these three LDA modules, each using a different set of features, the feature space was reduced from 12 to 3 features set. The LDA-SVM classifier utilizing the three most significant features set tested on the same 18 cohort patients data set obtained the sensitivity and specificity of 80% and 84% respectively. Moreover the SVM and LDA-SVM classifiers trained using 5 minutes EEG of 29 seizures recorded from 18 patients were validated on the entire EEG length of these 29 seizure recordings. The SVM and LDA-SVM classifiers scored the sensitivity (specificity) of 80% (86%) and 81% (85%) respectively. In addition to that, the SVM and LDA-SVM classifiers were tested on 6 seizures recorded from 4 more patients which are not involved in the system model development. Overall, among the SVM and LDA-SVM classifiers tested on 35 seizure recordings of 22 patients, the LDA-SVM classifier is found more suitable for automated detection system.

CHAPTER 6

DISCUSSION AND CONCLUSION

The seizure onset detection system plays an important role in the clinical environment to assist Neurologists in reviewing the long term video EEG recordings. The automated electrical seizure onset detection system identifies the nature of the EEG as normal or seizure episode. Moreover, the electrical onset detection would be helpful to develop a therapeutic intervention system for Anti Epileptic Drugs (AEDs) delivery to suppress the clinical seizure onset. In this thesis, Chapter 1 discussed the need for seizure detection system along with the basics of brain electrical activity, and measurement of brain potential using 10-20 electrode placement system. Furthermore, the types, electrical and clinical characteristics of the epileptic seizures were elaborated. A detailed literature survey of existing studies on seizure detection system was explored in Chapter 2. The statistical, time and frequency domain features extraction techniques used to determine the characteristics of electrical activity of the abnormal neurons producing recurrent seizures were elaborated. Likewise the machine learning algorithms adopted by the previous studies for automated seizure detection were discussed in Chapter 2.

6.1 Major outcomes of this research work

The majority of existing studies have included temporal and extra-temporal focal seizures to develop the automated seizure detection system (Aschenbrenner-Scheibe *et al.*, 2003; Meier *et al.*, 2008; Zandi *et al.*, 2010; Temko *et al.*, 2011; Rasekhi *et al.*, 2013; Bandarabadi *et al.*, 2015). Alternately, this thesis work restricted the investigation to detect the seizure originates from temporal lobe region. As discussed in Section. 1.4, EEG is used as the primary tool to capture the electrical activity of the brain. The previous studies used either intracranial or scalp recorded EEG to develop the seizure detection system. Compared to scalp-recorded EEG, intracranial EEG recordings are largely free of artifacts. However, as intracranial EEG is invasive, expensive, and undertaken on only in a minority of selected patients with AED-resistant epilepsy, with whom scalp-recorded EEG data have failed to provide the localization of seizure onset (Mormann *et al.*, 2005; Esteller *et al.*, 2005; Aarabi *et al.*, 2009; Blanco *et al.*, 2013). Moreover, most of the previous studies developed patient specific seizure detection system which needs the reconfiguration of the system for another individual (Shoeb *et al.*, 2004; Minasyan *et al.*, 2010; Chua *et al.*, 2011; Ozdemir and Yildirim, 2014; Selvakumari *et al.*, 2019). In this research work, utilizing 35 scalp recorded seizures of 22 temporal lobe epileptic patients from two local hospitals, the impact of sliding window length, features selection, normalization schemes, and the choice of machine learning algorithms were examined with the objective of patient independent system to detect the electrical onset of seizure.

In this work, the 35 seizures of 22 patients were divided into two sets. A set of 29 seizures recorded from 18 patients were used for the development of electrical seizure onset detection system. The electrical and clinical onset of 29 seizures were visually identified by the Neurologists. As discussed in Section. 3.1.1, 29 seizure recordings were truncated into 5 minutes EEG with respect to the clinical onset identified by the Neurologists. Each 5 minutes EEG consists of 2 minutes ictal EEG, and 3 minutes of preictal EEG preceding the ictal onset. The impact of sliding window length, features extraction and selection, normalization schemes and the machine learning algorithms for automated seizure detection were analyzed, using the 5 minutes EEG of these 29 seizure recordings. Another set of 6 seizure recordings from 4 patients were used for testing the system developed by using the 5 minutes EEG of 29 seizure recordings. The electrical onset of these 6 seizures visually identified by the Neurologists was correlated with the automated identification by the seizure detection system.

6.1.1 Sliding window length

The sliding window technique was used to extract the features from 5 minutes EEG of each seizure recording. The selection of optimal window length is the most crucial part of the seizure detection system. As discussed in Section. 4.2, optimal window length provides more meaningful features to capture the changes in EEG patterns during the seizure episode.

In previous studies many entropy estimators – Approximate Entropy (ApEn), Sample Entropy (SampEn), wavelet entropy and spectral entropy were applied to quantify the complexity of the brain to distinguish the normal and seizure activity (Kannathal *et al.*, 2005; Srinivasan *et al.*, 2007; Kumar *et al.*, 2010; Acharya *et al.*, 2012; Song *et al.*, 2012). The signal energy based measures were applied to detect large variations in energy distribution of recorded EEG during the pre-ictal, ictal and post-ictal phases (Litt *et al.*, 2001; Esteller *et al.*, 2005; Mormann *et al.*, 2005). Most of the studies on seizure detection have included statistical moments, band specific wavelet and spectral energies to capture the early changes in seizure episodes (Rasekhi *et al.*, 2013; Logesparan *et al.*, 2013; Moghim and Corne, 2014). The above presented studies calculated the entropy for different set of values for window length (N), embedding dimension (p) and filtering threshold (r) parameters. Moreover these studies used entropy based features to develop the classification model using scalp and intracranial EEG for normal and epileptic classes, respectively. Likewise the previous studies extracted signal energy, spectral and wavelet energy based features, and statistical moments of the intracranial and extracranial EEG for different length of overlapping and non-overlapping windows varying from 0.5 seconds to 1 hour (Paul, 2018). Hence comparing the performance of a classification model utilizing different set of features extracted from different length of windows is quite difficult.

In this research work, sliding windows of different length (M) of 2 seconds, 3 seconds, 4 seconds, and 5 seconds with M-1 seconds overlap with previous window were chosen to extract the features. As explained in Chapter 4, for each M seconds window a set of 22 time and frequency domain features were extracted from all 29 seizure recordings. As discussed in Section. 4.2, 2 seconds and 3 seconds windows showed more spiky profile in all the features which leads to the misclassification of normal and seizure condition. On the other hand 4 seconds, and 5 seconds windows produced smooth amplitude profile of the features which are more suitable to extract the characteristics of EEG. Hence in this research work, the 4 seconds window providing smooth features profiles was selected as optimal length for automated seizure detection.

6.1.2 Significant features selection

The time domain feature space consists of seven features namely mean, variance, skewness, kurtosis, approximate entropy (ApEn), sample entropy (SampEn) and signal energy. For all 29 seizure recordings, these seven features were extracted and the profile of these features were analyzed with respect to their electrical onset. Among the seven features, mean, skewness, kurtosis, approximate entropy (ApEn) and sample entropy (SampEn) have not shown consistent decrease or increase at the electrical onset of the recorded seizures. The other two features variance and signal energy showed consistent increase at the electrical onset of all recorded seizures. As explained in Section. 4.3.1, seven features samples extracted from each 5 minutes seizure recording were grouped into normal and seizure classes with respect to the electrical onset (Fig. 4.13). The separability of normal and seizure samples of all seven features were analyzed using Boxplot technique. Comparing the range of distribution of normal and seizure features samples shown in Fig. 4.14 and Table 4.1, variance and signal energy showed good separation in the order of 10 between the normal and seizure classes. Other five features namely mean, skewness, kurtosis, ApEn and SampEn showed very close values for the mean of normal and seizure classes. Specifically the distribution of normal and seizure samples of entropy features were completely overlapped as shown in Fig. 4.14. Hence, the signal energy equivalent to variance of DC component removed signal showing good separability among the normal and seizure classes were selected from the time domain features set.

Along with signal energy, wavelet and Fourier spectral energy features were extracted to capture the earliest rhythmic patterns produced by the hyper synchronized neurons. For all 29 seizure recordings, the profile of wavelet energy correlated with spectral energy profile in the selected frequency bands. Hence this work confined to simple Fourier spectral energies calculated in 1 to 25 Hz ($E_{lowband}$), 25 to 100 Hz ($E_{midband}$) and 100 to 200 Hz ($E_{highband}$) frequency bands. The spectral energy in 1 to 25 Hz band named as $E_{lowband}$ showed consistent increase at their electrical onset was selected for automated seizure detection. Among the different types of entropy mentioned in the previous section, the spectral entropy provides flexibility in frequency band selection for entropy estimation. Utilizing the flexibility in frequency band selection, the spectral entropy were estimated for different frequency bands selected within 1 to 200 Hz. From the analysis, the spectral entropy calculated for 3 to 12 Hz frequency band which correlates with temporal lobe ictal rhythms (Ebersole and Pacia, 1996), that show the large decrease at the electrical onset was chosen. The spectral entropy calculated for 3 to 12 Hz frequency band has not been reported in the previous studies on electrical seizure onset detection. The selected three significant features – spectral entropy, $E_{lowband}$ spectral energy and signal energy captured the electrical characteristics of temporal lobe seizure during the initial onset, spread to its neighbor regions and developed further into clinical seizure respectively.

While comparing the features selection, most of the previous studies utilizing frequency measures used standard spectral bands (0.5 to 4 Hz, 4 to 8 Hz, 8 to 13 Hz, 13 to 30 Hz, 30 to 48 Hz) to compute the spectral energy measures (Geng *et al.*, 2016; Minasyan *et al.*, 2010; Park *et al.*, 2011; Rasekhi *et al.*, 2013). Logesparan *et al.* (2013) analyzed the performance of 65 features and selected DWT based power in 0 to 3.125 Hz, 3.125 to 6.25 Hz, 6.25 to 12.5 Hz and 12.5 to 25 Hz frequency bands as the best discriminating features for seizure occurrence detection. In this thesis work, Fourier spectral energy in 1 to 25 Hz was selected as one of the best feature to train the classifiers for automated seizure electrical onset detection.

The feature extraction is the most significant part of any classification problem. However, due to the existence of high performance computing systems, in some of the recent studies feature extraction was not performed, and the deep learning models were trained using raw EEG signals (Acharya *et al.*, 2017; Hussein *et al.*, 2018; Zhou *et al.*, 2018; Zhao *et al.*, 2020). Acharya *et al.* (2017) proposed the first, 13 layer convolutional neural network (CNN) for the automated detection of epileptic seizure. The CNN model trained using small dataset classified the normal, preictal and ictal EEG samples of Bonn dataset with accuracy, specificity, and sensitivity of 88.67%, 90.00% and 95.00%, respectively. In another study, the original signals based on time or frequency domains were directly fed into the CNN trained using large data available from Freiburg and scalp CHB-MIT databases (Zhou *et al.*, 2018). Even though a CNN model learns the internal structure of data, and outperforms hand-crafted features selection techniques, the major constraint is the large number of learnable parameters, whose learning requires a huge volume of data. To overcome this limitation, One dimensional CNN which involves only fewer learnable parameters compared to standard CNN models, was used for the automated seizure detection (Ullah *et al.*, 2018).

The pyramid architecture based, 5 layers One dimensional CNN model proposed by Ullah et al. reduced the number of parameters significantly. Overall, the performance of SVM classifiers trained using hand crafted features are equal or better than CNN or one dimensional CNN models on Bonn dataset in identifying the normal and epileptic EEG samples (Sharma *et al.*, 2017; Faust *et al.*, 2010; Acharya *et al.*, 2011, 2012; Bhattacharyya *et al.*, 2017). The major advantage of CNN model is that separate steps of feature extraction not required to train the classifiers. On the other hand, the CNN model requires a large volume of data to obtain an optimum detection performance (Acharya *et al.*, 2017; Zhou *et al.*, 2018).

6.1.3 Patient dependent vs Patient independent system

Most of the systems proposed in previous studies were patient-dependent, and required expert input for channel selection and training data preparation to reconfigure the system for another patient (Shoeb *et al.*, 2004; Saab and Gotman, 2005; Minasyan *et al.*, 2010; Kharbouch *et al.*, 2011; Moghim and Corne, 2014; Rasekhi *et al.*, 2015). A generic SVM system using six ictal morphologies identified 91 seizures of 57 patients with a median detection latency of +1.6 seconds (ranging from -4 seconds to +10 seconds) and >96 % sensitivity. The system proposed by Meier *et al.* (2008) have included large number of electrodes to detect the onset of focal and secondarily generalized seizures from any part of the brain. Moreover, for each seizure recording the normalization was done in the features space by comparing the changes in feature samples extracted in short and long window of the same recording, and performance of SVM seizure detection was validated on leave one seizure out approach. Likewise a pediatric patient non-specific system developed by Orosco *et al.* (2016) used 22 electrodes to identify the seizure onset from any part of the brain. The Orosco's system using LDA classifier validated on leave one patient out approach to achieve the patient independentability.

Alternately, this thesis research work aimed to develop a self-sustaining seizure detection system by normalizing the EEG of all seizure recordings using a set of common normalization factors (Scheme 2). Using the average channel maximum and average channel minimum values computed from 29 seizures of 18 patients makes the EEG amplitude comparable across all the patients. The spectral entropy, $E_{lowband}$

spectral energy and signal energy features extracted from four channels T1, T3, T2, T4 of EEG were validated on leave one patient out approach to ensure the patient independent performance of the classifiers. In this thesis work the five classifiers LDA, NB, DT, SVM and KNN were trained for the automated detection of electrical onset of seizure. Each classifier identified the electrical onset of seizure at different latency compared to manual detection. With reference to Section. 5.2, the SVM performed better by detecting the electrical onset with the average sensitivity and specificity of 80% and 86% respectively. Another two classifiers DT and KNN scored the sensitivity (specificity) of 79% (80%) and 75% (85%) respectively. Other two classifiers LDA and NB performed with very low false positive rates and high missed seizure rate. These two classifiers scored the sensitivity (specificity) of 54% (94%) and 47% (96%) respectively. Further these classifiers were tested on cohort set containing more than one patient ensuring the patient independent performance. With reference to Section. 5.3 the SVM tested on cohort set secured the highest accuracy of 81% with the average sensitivity and specificity of 78% and 84% respectively. Since the SVM based seizure detection system performed better on individual and cohort patients' dataset, it is found more suitable for clinical routine.

The SVM model designed using 29 seizures of 18 patients were tested on their entire length of normal EEG recordings preceding the electrical onset of 29 seizures. The SVM classifier showed 86% of average specificity performance measure in detecting the normal samples of entire available length of EEG recordings (Table 5.7). Further improving the seizure detection performance as explained in Section. 5.4, the SVM classifier trained using the best feature (channel) of each features set – Spectral entropy in 3 to 12 Hz, $E_{lowband}$ spectral energy in 1 to 25 Hz and signal energy in 1 to 200 Hz extracted from four channels – selected through the LDA module obtained 85% of average specificity value. The performance of the SVM and LDA-SVM classifiers were also tested on six seizure recordings of 4 patients which are not involved in the system design. Overall, the SVM classifier identified the electrical onset of 31/35 seizures with the mean / median latency of 6.0 seconds / 2 seconds (ranging from -4 seconds to +32 seconds) and LDA-SVM classifier identified the onset of 30/35 seizures with the mean / median latency of 4.5 seconds / 1.5 seconds (ranging from -4 seconds to +32 seconds). Moreover, the computational complexity of SVM and LDA-SVM classifiers were analyzed. The SVM based seizure detection system needs normalizing factors and SVM classifier model to be stored in the memory. The other parameters computed for each 4 seconds window can be allocated dynamically during the run time of algorithm. In addition to the above memory resources requirement of the LDA-SVM based system needs three LDA models and their corresponding weight matrix for features space reduction of original test features samples to be stored in the memory. The computation of SVM and LDA-SVM based seizure detection algorithms simulated by using MATLAB running on i7 processor clocked at 2.7 GHz. The time lapse of SVM based algorithm starting from normalization to classification of given test features samples for a 4 seconds window is 0.17 seconds whereas the LDA-SVM based algorithm requires 0.21 seconds. Even though the LDA-SVM based seizure detection algorithm needs more computational resources and computation time, it is found more suitable for clinical study to detect the seizure onset little earlier than the SVM based detection system. The detection of seizure onset little earlier might be more useful for therapeutics interventions in managing the epilepsy.

6.1.4 Selection of critical parameters

In this thesis work, four parameters – EEG normalizing factors, selection of sliding window length, significant features set, and machine learning algorithms influencing the performance of seizure electrical onset detection system have been analyzed. Among these four parameters, the EEG normalizing factors – maximum and minimum values used for normalizing the EEG samples play an important role in determining the correct detection of normal and seizure features samples. In this research work, as discussed in Section. 3.3, two normalization schemes were used to design the automated system to detect the electrical onset of seizure.

The normalization is essential in patient independent seizure detection algorithm to correct large differences in the EEG amplitude across people, and in patient dependent algorithm to correct large amplitude variations over time due to the variations in brain function of the individuals (Logesparan *et al.*, 2011). The signal normalization module can be positioned within the seizure detection algorithm by three methods. The normalization factor is calculated on the emphasized signal characteristic (feature space) and applied to that feature itself (Casson and Rodriguez-Villegas, 2009; Minasyan *et al.*, 2010; Rasekhi *et al.*, 2013; Logesparan *et al.*, 2015). In another method, the normalization factor is calculated on the raw EEG signal $x(t)$ and applied to the raw EEG signal $x(t)$ Zhao *et al.* (2020). In the third method, the normalization factor is estimated from the raw EEG signal $x(t)$ and used to normalize the feature extracted from that signal $x(t)$ (Logesparan *et al.*, 2015). Logesparan *et al.* (2015) investigated the impact of five normalization techniques on seizure discrimination performance using the line length feature to emphasize seizure activity. Comparing five normalization schemes based upon the mean, median, standard deviation, signal peak and signal range, highlighted that the median decaying memory as the best approach for normalizing the line length features.

The scalp EEG amplitude varies widely across: different subjects and age groups; different electrode locations on the scalp; and over time due to changes in brain function or changes in the quality of the electrode connection to the scalp. Hence in this research work, adopting two normalizing factors computed from raw EEG signal, that EEG signal was normalized. The Scheme 1 normalizing technique was used to correct large amplitude variations over time due to the variations in brain function of the individuals whereas Scheme 2 normalizing technique was used to correct large differences in the EEG amplitude across people. The Scheme 1 SVM based seizure detection system evaluated on 29 seizure recordings had the average sensitivity and average specificity of 78% and 81% respectively. The Scheme 2 SVM based seizure detection system evaluated on 29 seizure recordings had the average sensitivity and average specificity of 80% and 86% respectively. Comparing the Scheme 1 and Scheme 2 based seizure detection system, the Scheme 2 based seizure detection system showed noticeable improvement in the classification performance.

According to the two sets of normalization factors derived from Scheme 1 and Scheme 2, the classifiers identified the electrical onset of 29 seizures with different detection latency. The Scheme 1 based SVM classifier identified the electrical onset of 25/29 seizures with median / mean detection latency of 4 seconds / 2 seconds (ranging from -20 seconds to +29 seconds) before their clinical onset. The Scheme 2 SVM classifier identified 25/29 seizures with median / mean latency of 2 seconds / 5.8 seconds (ranging from -4 seconds to +32 seconds). Further, considering the implementation of Scheme 1 based system for clinical study, obtaining the seizure specific maximum and minimum values to normalize the EEG of upcoming seizure is not possible. On the other hand, Scheme 2 based seizure detection system predetermined the maximum and minimum values used for normalizing the upcoming seizure EEG samples. Hence, Scheme 2 based automated system is found more suitable for clinical routine to detect the upcoming seizure onset. The Scheme 1 based system which needs the seizure specific maximum and minimum values for the normalization of EEG samples is not suitable to implement the patient independent seizure detection system. Alternatively, the normalization Scheme 2 used the common factors to normalize the EEG samples, is more suitable for the development of patient independent seizure detection system.

6.2 Conclusions

A patient independent system was designed to achieve the early detection of the electrical onset of temporal lobe epileptic seizures. The scalp-recorded EEG of 35 seizures recorded from 22 patients were used for this research work. The four channels T1, T3 and T2, T4 from left and right hemispheres of the brain were selected to pick-up the seizures originating from temporal lobe region. The 5 minutes EEG of 29 seizures recorded from 18 patients were selected to develop the seizure detection system. All the 29 seizure recordings were normalized using average maximum and average minimum computed from all 29 seizure recordings. The spectral entropy in 3 to 12 Hz, $E_{lowband}$ spectral energy in 1 to 25 Hz and signal energy in 1 to 200 Hz band – features were extracted for each 4 seconds window sliding over the 5 minutes EEG. The three features of four channels were used to train the LDA-SVM classifier for automated detection of the normal and seizure features samples. The performance of the LDA-SVM based system was validated on the long EEG recording of these 29 seizures. The LDA-SVM classifier obtained the average sensitivity and specificity of 81% and 85% respectively, on 29 seizure recordings arose from patient independent evaluation. Further the system was tested on six seizure recordings of four patients which are not involved in the model development. The LDA-SVM based system identified the electrical onset of 30/35 seizures with the median / mean latency of 1.5 seconds / 4.5 seconds (ranging from -4 seconds to +32 seconds).

The proposed patient independent system predefined the electrodes position to pick-up the seizures, and common normalizing factors which bring the EEG dynamics of all patients under one umbrella. Hence the proposed SVM model does not require experts' input to reconfigure the system for other individuals. This improves the comfort level of end users (Neurologists and Neurotechnologists). Such an automated seizure detection system would assist Neurologists in the visual analysis of long-term EEG recordings, and considerably reduce the time spent for reviewing the data. Automated seizure detection system can also help in delivering the stimulation in responsive neurostimulator system (RNS/closed loop interventions). If the system is trained with data from larger number of patients, the resulting patient-independent system might be of value in the presurgical evaluation of patients with drug-resistant Temporal Lobe Epilepsy (TLE). Moreover, the proposed seizure detection system would be helpful to inject the radio tracers for timing ictal Single Photon Emission Computed Tomography (SPECT) studies as well as for warning patients about upcoming clinical seizures.

6.3 Future scope of this work

The proposed SVM based system utilizing three features extracted from four channels identified the electrical onset of seizures with 80% (86%) of sensitivity (specificity) of identifying the given seizure (normal) EEG samples. Furthermore, as explained in Section. 5.4 the features of four channels were passed through LDA to select the significant features (channels) showing large discrimination among normal and seizure classes. The LDA-SVM classifier trained using the features of significant channels scored the sensitivity (specificity) of 81% (85%) of identifying the seizure (normal) EEG samples.

Analyzing the profile of interictal EEG patterns, it was noticed that the EEG recorded in routine clinical environment largely contaminated by head movement, body movement, eye movement and muscle/myogenic artifacts. If the electrodes are not placed properly, it picks-up the environmental potential due to the movement of caretakers. The presence of these movement and muscle artifacts in single channel or spreading over all the selected channels influence the false seizure detection by increasing the signal energy features fed to the classifier model. Apart from these potentials generated by non-brain activities, the brain produces few rhythmic intermittent patterns. These normal rhythmic patterns decrease the spectral entropy profile which leads to more false seizure detections as long as the patterns exists.

Incorporating a suitable algorithm for removing these artifacts without cutting the useful information for seizure onset detection, the false seizure detections could be reduced largely. Furthermore, improving the sensitivity and reducing the detection latency of the proposed system, the model parameters need to be optimized. The performance of the proposed system need to be improved in the above stated scenarios and further research may be focused to address these issues. Additionally, there is an opportunity in the future to test the performance of the proposed algorithm using CNN models for the detection of seizure onset.

REFERENCES

1. **Aarabi, A.**, **R. Fazel-Rezai**, and **Y. Aghakhani** (2009). A fuzzy rule-based system for epileptic seizure detection in intracranial EEG. *Clin Neurophysiol*, **120(9)**, 1648–57.

2. **Acharya, U.**, **S. Oh**, **Y. Hagiwara**, **J. Tan**, and **H. Adeli** (2017). Deep convolutional neural network for the automated detection and diagnosis of seizure using EEG signals. *Comput Biol Med*, **100**, 270–78.

3. **Acharya, U.**, **S. Sree**, **S. Chattopadhyay**, **W. Yu**, and **P. Ang** (2011). Application of recurrence quantification analysis for the automated identification of epileptic EEG signals. *International Journal of Neural Systems*, **21(3)**, 199–211.

4. **Acharya, U. R.**, **F. Molinari**, **S. S. Vinitha**, **S. Chattopadhyay**, **N. Kwan-Hoong**, and **J. S. Suri** (2012). Automated diagnosis of epileptic EEG using entropies. *Biomed Signal Process Control*, **7**, 401–08.

5. **Adeli, H.**, **Z. Zhou**, and **N. Dadmehr** (2003). Analysis of EEG records in an epileptic patient using wavelet transform. *J Neurosci Methods*, **123**, 69–87.

6. **Alarcon, G.**, **N. Kissani**, **M. Dad**, **R. D. Elwes**, **J. Ekanayake**, **M. J. Hennessy**, **M. Koutroumanidis**, **C. D. Binnie**, and **C. E. Polkey** (2001). Lateralizing and localizing values of ictal onset recorded on the scalp: evidence from simultaneous recordings with intracranial foramen ovale electrodes. *Epilepsia*, **42(11)**, 1426–37.

7. **Andrzejak, R. G.**, **K. Lehnertz**, **F. Mormann**, **C. Rieke**, **P. David**, and **C. E. Elger** (2001). Indications of nonlinear deterministic and finite-dimensional structures in time series of brain electrical activity: Dependence on recording region and brain state. *Phys Rev E Stat Nonlin Soft Matter Phys*, **64**, 061907.

8. **Aschenbrenner-Scheibe, R.**, **T. Maiwald**, **M. Winterhalder**, **H. U. Voss**, **J. Timmer**, and **A. Schulze-Bonhage** (2003). How well can epileptic seizures be predicted? an evaluation of nonlinear method. *Brain*, **126**, 2616–26.

9. **Baldominos, A.** and **C. Ramon-Lozano**, Optimizing EEG Energy-based Seizure Detection using Genetic Algorithms. *In 2017 IEEE Congress on Evolutionary Computation (CEC), 2017, 2338–45*. IEEE, 2017.

10. **Bandarabadi, M.**, **C. Teixeira**, **T. Netoff**, **K. Parhi**, and **A. Dourado**, Robust and low complexity algorithms for seizure detection. *In The 36th IEEE Annual International Conference on Engineering in Medicine and Biology Society, Chicago, IL; 2014, 4447–50*. IEEE, 2014.

11. **Bandarabadi, M.**, **C. A. Teixeira**, **J. Rasekhi**, and **A. Dourado** (2015). Epileptic seizure prediction using relative spectral power features. *Clin Neurophysiol*, **126(2)**, 237–48.

12. **Bhattacharyya, A.**, **R. Pachori**, **A. Upadhyay**, and **U. Acharya** (2017). Tunable-Q wavelet transform based multiscale entropy measure for automated classification of

epileptic EEG signals. *Application of Signal Processing Methods for Systematic Analysis of Physiological Health*, **7(4)**, 385.

13. **Bhople, A.** (2012). Fast Fourier transform based classification of epileptic seizure using artificial neural network. *Int J Adv Res Comput Sci Softw Eng*, **2(4)**.

14. **Blanco, S., A. Garay**, and **D. Coulombie** (2013). Comparison of frequency bands using spectral entropy for epileptic seizure prediction. *ISRN Neurol*, **2013**, 1–5.

15. **Bogaarts, J., E. Gommer, D. Hilkman, V. van Kranen-Mastenbroek**, and **J. Reulen** (2014). Eeg feature pre-processing for neonatal epileptic seizure detection. *Ann Biomed Eng*, **42(11)**, 2360–68.

16. **Bogaarts, J., E. Gommer, D. Hilkman, V. van Kranen-Mastenbroek**, and **J. Reulen** (2016). Improved epileptic seizure detection combining dynamic feature normalization with EEG novelty detection. *Med Biol Eng Comput*, **54(12)**, 1883–92.

17. **Brekelmans, G. J. F., W. Van Emde Boas, D. N. Velis, A. C. Van Huffelen, R. M. C. Debets**, and **C. W. M. Van Veelen** (1995). Mesial temporal versus neocortical temporal lobe seizures: Demonstration of different electroencephalographic spreading patterns by combined use of subdural and intracerebral electrodes. *J. Epilepsy*, **8**, 309–20.

18. **Burgerman, R. S., M. R. Sperling, J. A. French, A. J. Saykin**, and **M. J. O'Connor** (1995). Comparison of Mesial Versus neocortical Onset Temporal Lobe Seizures: Neurodiagnostic Findings and Surgical Outcome. *Epilepsia*, **36(7)**, 662–70.

19. **Camfield, P.** and **C. Camfield** (2010). Idiopathic generalized epilepsy with generalized tonic-clonic seizures (IGE-GTC): A population-based cohort with 20 year follow up for medical and social outcome. *Epilepsy Behav*, **18**, 61–3.

20. **Carney, P. R., S. Myers**, and **J. D. Geyer** (2011). Seizure prediction: Methods. *Epilepsy Behav*, **22**, S94–101.

21. **Casson, A. J.** and **E. Rodriguez-Villegas** (2009). Toward online data reduction for portable electroencephalography systems in epilepsy. *IEEE TRANSACTIONS ON BIOMEDICAL ENGINEERING*, **56(12)**, 2816–25.

22. **Cherian, A., A. Radhakrishnan, S. Parameswaran, R. Varma**, and **K. Radhakrishnan** (2012). Do sphenoidal electrodes aid in surgical decision making in drug resistant temporal lobe epilepsy? *Clin Neurophysiol*, **123**(3), 463–70.

23. **Choe, S. H., Y. G. Chung**, and **S. P. Kim**, Statistical spectral feature extraction for classification of epileptic EEG signals. *In International Conference on Machine Learning and Cybernetics, Qingdao, China; July 2010 3180–5*. IEEE, 2010.

24. **Chua, E. C., K. Patel, M. Fitzsimons**, and **C. J. Bleakley** (2011). Improved patient specific seizure detection during the pre-surgical evaluation. *Clin Neurophysiol*, **122**, 672–9.

25. **Dericioglu, N.** and **S. Saygi** (2008). Ictal scalp eeg findings in patients with mesial temporal lobe epilepsy. *Clin EEG Neurosci*, **39(1)**, 20–7.

26. **Duun-Henriksen, J., T. Kjaer, R. Madsen, L. Remvig, C. Thomsen**, and **H. Sorensen** (2012). Channel selection for automatic seizure detection. *Clin Neurophysiol*, **123**, 84–92.

27. **Ebersole, J. S.** and **S. V. Pacia** (1996). Localization of temporal lobe foci by ictal EEG patterns. *Epilepsia*, **37**(4), 386–99.

28. **Emami, A.**, **N. Kunii**, **T. Matsuo**, **T. Shinozaki**, **K. Kawaie**, and **H. Takahashia** (2019). Seizure detection by convolutional neural network-based analysis of scalp electroencephalography plot images. *NeuroImage: Clinical*, **22**(101684).

29. **Engel, J. J.** (2001). Mesial Temporal Lobe Epilepsy: What Have we Learned? *Neuroscientist*, **7**, 340–52.

30. **Engel, J. J.** and **R. F. Ackermann** (1980). Interictal EEG spikes correlate with decreased, rather than increased, epileptogenicity in amygdaloid kindled rats. *Brain Research*, **190**, 543–48.

31. **Esteller, R.**, **J. Echauz**, **M. D'Alessandro**, and **G. Worrell** (2005). Continuous energy variation during the seizure cycle: towards an on-line accumulated energy. *Clin Neurophysiol*, **3**(116), 517–26.

32. **Exarchos, T. P.**, **A. T. Tzallas**, **D. I. Fotiadis**, **S. Konitsiotis**, and **S. Giannopoulos** (2006). EEG transient event Detection and Classification Using Association Rules. *IEEE Trans Inf Technol Biomed*, **10**(3), 451–7.

33. **F. Angeleri, F.**, **S. Giaquinto**, and **G. F. Marchesi** (1972). Temporal Distribution of Interictal and Ictal Discharges from Penicillin Foci in Cats. *In Petsche H, Brazier MAB (eds): Synchronization of EEG Activity in Epilepsies. New York, Springer*, 221–34.

34. **Falco-Walter, J. J.**, **I. E. Scheffer**, and **R. S. Fisher** (2018). The new definition and classification of seizures and epilepsy. *Epilepsy Res*, **139**, 73–9.

35. **Faust, O.**, **U. Acharya**, **C. Lim**, and **B. Sputh** (2010). Automatic identification of epileptic and background EEG signals using frequency domain parameters. *International Journal of Neural Systems*, **20**(2), 159–76.

36. **Fisher, R. S.**, **H. E. Scharfman**, and **M. deCurtis** (2014). How can we identify ictal and interictal abnormal activity?. *Adv Exp Med Biol*, **813**, 71–80.

37. **Fried, I.** (1997). Auras and experiential responses arising in the temporal lobe. *J Neuropsychiatry Clin Neurosci*, **9**(3), 420–8.

38. **Fukami, T.**, **T. Shimada**, and **B. Ishikawa** (2018). Fast EEG spike detection via eigenvalue analysis and clustering of spatial amplitude distribution. *J Neural Eng*, **15**(3).

39. **Furbass, F.**, **S. Kampusch**, **E. Kaniusas**, **J. Koren**, **S. Pirker**, **R. Hopfengartner**, **H. Stefan**, **T. Kluge**, and **C. Baumgartner** (2017). Automatic multimodal detection for long-term seizure documentation in epilepsy. *Clin Neurophysiol*, **128**(8), 1466–72.

40. **Gabor, A. J.** (1998). Seizure detection using a self-organizing neural network: validation and comparison with other detection strategies. *Electroencephalogr Clin Neurophysiol*, **107**, 27–32.

41. **Gabor, A. J.** and **M. Seyal** (1992). Automated interictal EEG spike detection using artificial neural networks. *Electroencephalogr Clin Neurophysiol*, **83**, 271–80.

42. **Gandhi, T., P. Chakraborty, G. Roy**, and **B. Panigrahi** (2012). Discrete harmony search based expert model for epileptic seizure detection in electroencephalography. *Expert Syst Appl*, **39**, 4055–62.

43. **Gandhi, T., B. Panigrahi, M. Bhatia**, and **S. Anand** (2010). Expert model for detection of epileptic activity in eeg signature. *Expert Syst Appl*, **37**, 3513–20.

44. **Geng, D., W. Zhou, Y. Zhang**, and **S. Geng** (2016). Epileptic seizure detection based on improved wavelet neural networks in long-term intracranial EEG. *Biocyber Biomed Eng*, **36(2)**, 375–84.

45. **Gigola, S., F. Ortiz, C. E. D'Attellis, W. Silva**, and **S. Kochen** (2004). Prediction of epileptic seizures using accumulated energy in a multiresolution framework. *J Neurosci Methods*, **138(1-2)**, 107–11.

46. **Gotman, J.** (1982). Automatic recognition of epileptic seizures in the EEG. *Electroencephalogr Clin Neurophysiol*, **54(5)**, 530–40.

47. **Gotman, J.** (1999). Automatic detection of seizures and spikes. *J Clin Neurophysiol*, **16(2)**, 130–40.

48. **Gotman, J., J. Ives, P. Gloor, A. Olivier**, and **L. F. Quesney** (1982). Changes in interictal EEG spiking and seizure occurrence in humans (Abstr.). *Epilepsia*, **23**, 432–33.

49. **Gotman, J.** and **M. G. Marciani** (1985). Electroencephalographic spiking activity, drug levels, and seizure occurrence in epileptic patients. *Ann Neurol*, **17**(6), 597–603.

50. **Greene, B. R., S. Faula, W. P. Marnanea, G. Lightbody, I. Korotchikova**, and **G. B. Boylan** (2008). A comparison of quantitative eeg features for neonatal seizure detection. *Clin Neurophysiol*, **119**, 1248–61.

51. **Hamad, A., E. H. Houssein, A. E. Hassanien**, and **A. A. Fahmy**, A Hybrid EEG signals classification approach based on Grey Wolf optimizer enhanced SVMs for epileptic detection. *In The International Conference on Advanced Intelligent Systems and Informatics, Cairo, Egypt; Sep 2017 108–17*. Springer, 2017.

52. **Hamer, H. M., I. Najm, A. Mohamed**, and **E. Wyllie** (1999). Interictal epileptiform discharges in temporal lobe epilepsy due to hippocampal sclerosis versus medial temporal lobe tumors. *Epilepsia*, **40(9)**, 1261–8.

53. **Hartmann, M., F. Furbass, H. Perko, A. Skupch, K. Lackmayer, C. Baumgartner**, and **T. Kluge**, EpiScan: online seizure detection for epilepsy monitoring units. *In Annual International Conference of the IEEE Engineering in Medicine and Biology Society, 2011 6096–99*. IEEE, 2011.

54. **Hendelman, W., P. Humphreys**, and **C. R. Skinner**, *The Integrated Nervous System, A Systematic Diagnostic Case-Based Approach, Second Edition*. CRC Press,Taylor and Francis Group, 2017.

55. **Hernández, D., L. Trujillo, E. Z-Flores, O. Villanueva**, and **O. Romo-Fewell** (). Detecting Epilepsy in EEG Signals Using Time, Frequency and Time-Frequency Domain Features. In: Sanchez M., Aguilar L., Castañón-Puga M., Rodríguez-Díaz A. (eds) Computer Science and EngineeringÜTheory and Applications. Studies in Systems, Decision and Control, vol 143. Springer, Cham.

56. **Hu, W.**, **J. Cao**, **X. Lai**, and **J. Liu** (2019). Mean amplitude spectrum based epileptic state classification for seizure prediction using convolutional neural networks. *Journal of Ambient Intelligence and Humanized Computing*.

57. **Hunyadi, B.**, **M. Signoretto**, **W. Van Paesschen**, **J. A. Suykens**, **S. Van Huffel**, and **M. De Vos** (2012). Incorporating structural information from the multichannel EEG improves patient-specific seizure detection. *Clin Neurophysiol*, **123(12)**, 2352–61.

58. **Hussein, R.**, **H. Palangi**, **R. Ward**, and **Z. Wang** (2018). Epileptic seizure detection: a deep learning approach.

59. **Inan, Z. H.** and **M. Kuntalp** (2007). A study on fuzzy C-means clustering-based systems in automatic spike detection. *Comput Biol Med*, **37**, 1160–6.

60. **Indiradevi, K. P.**, **E. Elias**, **P. S. Sathidevi**, **S. Dinesh Nayak**, and **K. Radhakrishnan** (2008). A multilevel wavelet approach for automatic detection of epileptic spikes in the electroencephalogram. *Comput Biol Med*, **38**(7), 805–16.

61. **Iscan, Z.**, **D. Zmray**, and **D. Tamer** (2011). Classification of electroencephalogram signals with combined time and frequency features. *Expert Syst. Appl*, **38(8)**, 10499–505.

62. **Jory, C.**, **R. Shankar**, **D. Coker**, **B. McLean**, **J. Hanna**, and **C. Newman** (2016). Safe and sound? A systematic literature review of seizure detection methods for personal use. *Seizure*, **36**, 4–15.

63. **Julien, R.**, **C. Advokar**, and **J. Comaty**, *A Primer of Drug Action: A Comprehensive Guide to the Actions, Uses, and Side Effects of Psychoactive Drugs*. New York: Worth Publishers; 2008, 2008.

64. **Kafashan, M.**, **S. Ryu**, and **M. J. Hargis et al.** (2017). EEG dynamical correlates of focal and diffuse causes of coma. *BMC Neurol*, **17(1)**, 197.

65. **Kamath, C.** (2013). A new approach to detect epileptic seizures in electroencephalograms using teager energy. *ISRN Biomedical Engineering*, **2013**, 1–14.

66. **Kannathal, N.**, **M. L. Choo**, **U. R. Acharya**, and **P. K. Sadasivan** (2005). Entropies for detection of epilepsy in EEG. *Comput Methods Programs Biomed*, **80**(3), 187–94.

67. **Kelly, K. M.**, **D. S. Shiau**, **R. T. Kern**, **J. H. Chien**, **M. C. Yang**, **K. A. Yandora**, **J. P. Valeriano**, **J. J. Halford**, and **J. C. Sackellares** (2010). Assessment of a scalp EEG-based automated seizure detection system. *Clin Neurophysiol*, **121(11)**, 1832–4.

68. **Khamis, H.**, **A. Mohamed**, and **S. Simpson** (2009). Seizure state detection of temporal lobe seizures by autoregressive spectral analysis of scalp EEG. *Clin Neurophysiol*, **120(8)**, 1479–88.

69. **Khamis, H.**, **A. Mohamed**, and **S. Simpson** (2013). FrequencyŰmoment signatures: A method for automated seizure detection from scalp EEG. *Clin Neurophysiol*, **124(12)**, 2317–27.

70. **Kharbouch, A.**, **A. Shoeb**, **J. Guttag**, and **S. S. Cash** (2011). An algorithm for seizure onset detection using intracranial EEG. *Epilepsy Behav*, **22(1)**, S29–S35.

71. **Ko, C. W.** and **H. W. Chung** (2000). Automatic spike detection via an artificial neural network using raw eeg data: effects of data preparation and implications in the limitations of online recognition. *Clin Neurophysiol*, **111**(3), 477–81.

72. **Kononenko, I.**, Estimating attributes: analysis and extensions of RELIEF. *In In European conference on machine learning, Berlin, Heidelberg; 1994, 171–182.* Springer, 1994.

73. **Kononenko, I.**, **E. Simec**, and **M. Robnik-Sikonja** (1997). Overcoming the myopia of inductive learning algorithms with RELIEFF. *Applied Intelligence*, **7**(1), 39–55.

74. **Kumar, S. P.**, **N. Sriram**, **P. G. Benakop**, and **B. C. Jinaga** (2010). Entropies based detection of epileptic seizures with artificial neural network classifiers. *Expert Syst Appl*, **37**(2), 3284–91.

75. **Lange, H. H.**, **J. P. Lieb**, **J. J. Engel**, and **P. H. Crandall** (1983). Temporo-spatial patterns of pre-ictal spike activity in human temporal lobe epilepsy. *Electroencephalogr Clin Neurophysiol*, **56**, 543–55.

76. **Le Van Quyen, M.**, **J. Martinerie**, **V. Navarro**, **P. Boon**, **M. D'Have**, **C. Adam**, **B. Renault**, **F. Varela**, and **M. Baulac** (2001). Anticipation of epileptic seizures from standard EEG recordings. *Lancet*, **357**(9251), 183–8.

77. **Li, P.**, **C. Karmakar**, **J. Yearwood**, **S. Venkatesh**, **M. Palaniswami**, and **C. Liu** (2018). Detection of epileptic seizure based on entropy analysis of short-term EEG. *PLoS ONE*, **13**(3), e0193691.

78. **Lieb, J. P.**, **S. C. Woods**, **A. Siccardi**, **P. H. Crandall**, **D. O. Walter**, and **B. Leake** (1978). Quantitative analysis of depth spiking in relation to seizure foci in patients with temporal lobe epilepsy. *Electroencephalogr Clin Neurophysiol*, **44**(5), 641–63.

79. **Litt, B.** and **J. Echauz** (2002). Prediction of epileptic seizures. *Lancet Neurol*, **1**(1), 22–30.

80. **Litt, B.**, **R. Esteller**, **J. Echauz**, **M. D'Alessandro**, **R. Shor**, **T. Henry**, **P. Pennell**, **C. Epstein**, **R. Bakay**, **M. Dichter**, and **G. Vachtsevanos** (2001). Epileptic seizures May begin Hours in advance of Clinical onset: A report of Five patients. *Neuron*, **30**(1), 51–64.

81. **Logesparan, L.**, **A. J. Casson**, **S. A. Imtiaz**, and **E. Rodriguez-Villegas**, Discriminating between best performing features for seizure detection and data selection. *In The 35th IEEE Annual International Conference on Engineering in Medicine and Biology Society, Osaka, Japan; July 2013 1692–5.* IEEE, 2013.

82. **Logesparan, L.**, **A. J. Casson**, and **E. Rodriguez-Villegas**, Assessing the impact of signal normalization: preliminary results on epileptic seizure detection. *In Annu Int Conf IEEE Eng Med Biol Soc, Boston; 2011 1439–42.* IEEE, 2011.

83. **Logesparan, L.**, **A. J. Casson**, and **E. Rodriguez-Villegas** (2012). Optimal features for online seizure detection. *Med Biol Eng Comput*, **50**(7), 659–69.

84. **Logesparan, L.**, **E. Rodriguez-Villegas**, and **A. J. Casson** (2015). The impact of signal normalization on seizure detection using line length features. *Med Biol Eng Comput*, **53**, 929–42.

85. **Maiwald, T., M. Winterhalder, R. Aschenbrenner-Scheibe, H. U. Voss, A. Schulze-Bonhage,** and **J. Timmer** (2004). Comparison of three nonlinear seizure prediction methods by means of the seizure prediction characteristic. *Physica D*, **194**, 357–68.

86. **Malmivuo, J.** and **R. Plonsey**, *Bioelectromagnetism*. New York: Oxford University Press, 1995.

87. **McSharry, P. E., L. A. Smith,** and **L. Tarassenko** (2003). Comparison of Predictability of Epileptic Seizures by a Linear and a Nonlinear method. *IEEE Trans Biomed Eng*, **50**(5), 628–33.

88. **Meier, R., H. Dittrich, A. Schulze-Bonhage,** and **A. Aertsen** (2008). Detecting epileptic seizures in long-term human EEG: a new approach to automatic online and real-time detection and classification of polymorphic seizure patterns. *J Clin Neurophysiol*, **25**(3), 119–31.

89. **Minasyan, G. R., J. B. Chatten, M. J. Chatten,** and **R. N. Harner** (2010). Patient-Specific Early Seizure Detection from Scalp EEG. *J Clin Neurophysiol*, **27**(3), 163Ű–78.

90. **Moghim, N.** and **D. W. Corne** (2014). Predicting epileptic seizures in advance. *PLoS One*, **9**(6), e99334.

91. **Mormann, F., T. Kreuz, C. Rieke, R. G. Andrzejak, A. Kraskov, P. David, C. E. Elger,** and **K. Lehnertz** (2005). On the predictability of epileptic seizures. *Clin Neurophysiol*, **116**(3), 569–87.

92. **Nishida, S., M. Nakamura, A. Ikeda,** and **H. Shibasaki** (1999). Signal separation of background EEG and spike by using morphological filter. *Med Eng Phys*, **21**, 601–8.

93. **Ord, J.**, *Families of frequency distributions*. London: London Publishers; 1972, 1972.

94. **Orosco, L., A. G. Correa, P. Diez,** and **E. Laciar** (2016). Patient non-specific algorithm for seizures detection in scalp EEG. *Comput Biol Med*, **71**, 128–34.

95. **Osorio, I., M. G. Frei,** and **S. B. Wilkinson** (1998). Real-time automated detection and quantitative analysis of seizures and short-term prediction of clinical onset. *Epilepsia*, **39**(6), 615–27.

96. **Ozdamar, O.** and **T. Kalayci** (1998). Detection of spikes with artificial neural networks using raw EEG. *Comput Biomed Res*, **31**(2), 122–42.

97. **Ozdemir, N.** and **E. Yildirim** (2014). Patient specific seizure prediction system using H ilbert spectrum and B ayesian networks classifiers. *Comput Math Methods Med*, **2014**.

98. **Park, Y., L. Luo, K. K. Parhi,** and **T. Netoff** (2011). Seizure prediction with spectral power of EEG using cost-sensitive support vector machines. *Epilepsia*, **52**(10), 1761–70.

99. **Paul, Y.** (2018). Various epileptic seizure detection techniques using biomedical signals: a review. *Brain Inf*, **5**(6), 1–19.

100. **Pfander, M., S. Arnold, A. Henkel, S. Weil, K. J. Werhahn, I. Eisensehr, P. A. Winkler**, and **S. Noachtar** (2002). Clinical features and EEG findings differentiating mesial from neocortical temporal lobe epilepsy. *Epileptic Disord*, **4(3)**, 189–95.

101. **Pincus, S. M.** (1991). Approximate entropy as a measure of system complexity. *Proc Natl Acad Sci USA*, **88**(6), 2297–301.

102. **Polat, K.** and **S. Gunes** (2007). Classification of epileptiform eeg using a hybrid system based on decision tree classifier and fast Fourier transform. *Appl Math Comput*, **187**, 1017–26.

103. **Pradhan, N., P. K. Sadasivan**, and **G. R. Arunodaya** (1996). Detection of seizure activity in EEG by an artificial neural network: a preliminary study. *Comput Biomed Res*, **29**, 303–13.

104. **Qu, H.** and **J. Gotman** (1995). A seizure warning system for long-term epilepsy monitoring. *Neurology*, **45(12)**, 2250–4.

105. **Quintero-Rincon, A., M. Pereyra, C. D'Giano, H. Batatia**, and **M. Risk** (2016). A new algorithm for epilepsy seizure onset detection and spread estimation from EEG signals. *Journal of Physics: Conference Series*, **705**(1), 012032.

106. **Ramgopal, S., S. Thome-Souza, M. Jackson, N. Kadish, I. Sanchez Fernandez, J. Klehm, W. Bosl, C. Reinsberger, S. Schachter**, and **T. Loddenkemper** (2014). Seizure detection, seizure prediction, and closed-loop warning systems in epilepsy. *Epilepsy Behav*, **37**, 291–307.

107. **Rasekhi, J., M. R. Mollaei, M. Bandarabadi, C. A. Teixeira**, and **A. Dourado** (2013). Preprocessing effects of 22 linear univariate features on the performance of seizure prediction methods. *J Neurosci Methods*, **217**(1-2), 9–16.

108. **Rasekhi, J., M. R. Mollaei, M. Bandarabadi, C. A. Teixeira**, and **A. Dourado** (2015). Epileptic Seizure Prediction based on Ratio and Differential Linear Univariate Features. *J Med Signals Sens*, **5**(1), 1–11.

109. **Rathore, C., C. Kesavadas, J. Ajith, A. Sasikala, P. Sarma**, and **K. Radhakrishnan** (2011). Cost-effective utilization of single photon emission computed tomography (SPECT) in decision making for epilepsy surgery. *Seizure*, **20(2)**, 107–14.

110. **Ravish, D. K., S. S. Devi**, and **S. G. Krishnamoorthy** (2013). Detection of Epileptic Seizure in EEG recordings by Spectral Method and Statistical Analysis. *Journal of Applied Sciences*, **13**, 207–19.

111. **Richman, J. S.** and **J. R. Moorman** (2000). Physiological time-series analysis using approximate entropy and sample entropy. *Am J Physiol Heart Circ Physiol*, **278**, H2039–49.

112. **Rogowski, Z., I. Gath**, and **E. Bental** (1981). On the prediction of epileptic seizures. *Biol Cybern*, **42**(9-15).

113. **Saab, M. E.** and **J. Gotman** (2005). A system to detect the onset of epileptic seizures in scalp EEG. *Clin Neurophysiol*, **116**, 427Ű–42.

114. **Selvakumari, R. S.**, **M. Mahalakshmi**, and **P. Prashalee** (2019). Patient-Specific Seizure Detection Method using Hybrid Classifier with Optimized Electrodes. *J Med Syst*, **43(5)**, 121.

115. **Selvitelli, M. F.**, **L. M. Walker**, **D. L. Schomer**, and **B. S. Chang** (2010). The relationship of interictal epileptiform discharges to clinical epilepsy severity: A study of routine EEGs and review of the literature. *J Clin Neurophysiol*, **27(2)**, 87–92.

116. **Sharaf, A. I.**, **M. A. El-Soud**, and **I. M. El-Henawy** (2018). An Automated Approach for Epilepsy Detection Based on Tunable Q-Wavelet and Firefly Feature Selection Algorithm. *Int J Biomed Imaging*.

117. **Sharma, M.**, **R. Pachori**, and **U. Acharya** (2017). A new approach to characterize epileptic seizures using analytic time-frequency flexible wavelet transform and fractal dimension. *Pattern Recognition Letters.*, **94**, 172–79.

118. **Sharmila, A.** and **P. Geethanjali** (2019). A review on the pattern detection methods for epilepsy seizure detection from EEG signals. *Biomed Eng Biomed Tech*.

119. **Shen, T. W.**, **X. Kuo**, and **Y. L. Hsin**, Ant K-means Clustering Method on Epileptic Spike Detection. *In Fifth International Conference on Natural Computation, Tianjin, China; Aug 2009 334–8*. IEEE, 2009.

120. **Shiau, D. S.**, **J. J. Halford**, **K. M. Kelly**, **R. T. Kern**, **M. Inman**, **J. H. Chien**, **P. M. Pardalos**, **M. C. K. Yang**, and **J. C. Sackellares** (2010). Singularity-based automated seizure detection system for scalp EEG monitoring. *Cybern Syst Anal*, **46(6)**, 922–35.

121. **Shoeb, A.** (2009). Application of machine learning to epileptic seizure onset detection and treatment. *Ph.D. thesis, Massachusetts Institute of Technology*.

122. **Shoeb, A.**, **H. Edwards**, **J. Connolly**, **B. Bourgeois**, **S. T. Treves**, and **J. Guttag** (2004). Patient-specific seizure onset detection. *Epilepsy Behav*, **5**, 483–98.

123. **Shoeb, A.** and **J. Guttag**, Application of machine learning to epileptic seizure detection. *In The 27th International Conference on Machine Learning , Haifa, Israel; June 2010 975–82*. 2010.

124. **Sitt, J. D.**, **J. R. King**, **I. El Karoui**, **B. Rohaut**, **F. Faugeras**, **A. Gramfort**, **L. Cohen**, **M. Sigman**, **S. Dehaene**, and **L. Naccache** (2014). Large scale screening of neural signatures of consciousness in patients in a vegetative or minimally conscious state. *Brain*, **137(Pt 8)**, 2258–70.

125. **Song, Y.**, **J. Crowcroft**, and **J. Zhang** (2012). Automatic epileptic seizure detection in eegs based on optimized sample entropy and extreme learning machine. *J Neurosci Methods*, **210**(2), 132–46.

126. **Srinivasan, V.**, **C. Eswaran**, and **N. Sriraam** (2007). Approximate entropy-based epileptic EEG detection using artificial neural networks. *IEEE Trans Inf Technol Biomed*, **11**(3), 288–95.

127. **Stevens, J. R.**, **B. L. Lonsbury**, and **S. L. Goel** (1972). Seizure occurrence and interspike interval: telemetered electroencephalogram studies. *Arch. Neurol. (Chic.)*, **26**, 409–19.

128. **Subasi, A.** (2006). Automatic detection of epileptic seizure using dynamic fuzzy neural networks. *Expert Syst Appl*, **31(2)**, 320–8.

129. **Subasi, A., J. Kevric,** and **M. Abdullah Canbaz** (2019). Epileptic seizure detection using hybrid machine learning methods. *Neural Comput Applic*, **31**, 317–25.

130. **Swami, P., M. Bhatia, M. Tripathi, P. Sarat Chandra, B. Panigrahi,** and **T. Gandhi** (2019). Selection of optimum frequency bands for detection of epileptiform patterns. *Healthcare Technology Letters*, **6(5)**, 126–31.

131. **Swami, P., T. Gandhi, B. Panigrahi, M. Tripathi,** and **S. Anand** (2016). A novel robust diagnostic model to detect seizures in electroencephalography. *Expert Syst Appl*, **56**, 116–30.

132. **Temko, A., E. Thomas, G. Boylan, W. Marnane,** and **G. Lightbody**, An SVM-based system and Its Performance for Detection of Seizures in Neonates. *In Annual International Conference of the IEEE Engineering in Medicine and Biology Society, Minnesota, USA; Sep 2009 2643–6.* IEEE, 2009.

133. **Temko, A., E. Thomas, G. Boylan, W. Marnane,** and **G. Lightbody** (2015). Clinical implementation of a neonatal seizure detection algorithm. *Decis Support Syst*, **70**, 86–96.

134. **Temko, A., E. Thomas, W. Marnane, G. Lightbody,** and **G. Boylan** (2011). EEG-based neonatal seizure detection with Support Vector Machines. *Clin Neurophysiol*, **122**, 464Ű–73.

135. **Ulate-Campos, A., F. Coughlin, M. Gainza-Lein, I. S. Fernandez, P. L. Pearl,** and **T. Loddenkemper** (2016). Automated seizure detection systems and their effectiveness for each type of seizure. *Seizure*, **40**, 88–101.

136. **Ullah, I., M. Hussain, E. Qazi,** and **H. Aboalsamh** (2018). An automated system for epilepsy detection using EEG brain signals based on deep learning approach. *Expert Syst Appl*, **107**, 61–71.

137. **Van Putten, M. J.** (2007). The revised brain symmetry index. *Clin Neurophysiol*, **118(11)**, 2362–7.

138. **Van Putten, M. J., T. Kind, F. Visser,** and **V. Lagerburg** (2005). Detecting temporal lobe seizures from scalp EEG recordings: a comparison of various features. *Clin Neurophysiol*, **116(10)**, 2480–9.

139. **Wang, L., W. Xue, Y. Li, M. Luo, J. Huang, W. Cui,** and **C. Huang** (2017). Automatic Epileptic Seizure Detection in EEG Signals Using Multi-Domain Feature Extraction and Nonlinear analysis. *Entropy*, **19**(6), 1–17.

140. **Webber, W. R., R. P. Lesser, R. T. Richardson,** and **K. Wilson** (1996). An approach to seizure detection using an artificial neural network (ANN). *Electroencephalogr Clin Neurophysiol*, **98**(9-15), 250–70.

141. **Weng, W.** and **K. Khorasani** (1996). An Adaptive Structure Neural Networks with Application to EEG Automatic Seizure Detection. *Neural Networks*, **9(7)**, 1223–40.

142. **Wilson, S. B., M. L. Scheuer, R. G. Emerson,** and **A. J. Gabor** (2004). Seizure detection: evaluation of the Reveal algorithm. *Clin Neurophysiol*, 2280–91.

143. **Winterhalder, M., T. Maiwald, H. U. Voss, R. Aschenbrenner-Scheibe, J. Timmer,** and **A. Schulze-Bonhage** (2003). The seizure prediction characteristic: a general framework to assess and compare seizure prediction methods. *Epilepsy Behav*, **4**(3), 318–25.

144. **Wu, L.** and **J. Gotman** (1998). Segmentation and classification of EEG during epileptic seizures. *Electroencephalogr Clin Neurophysiol*, **106**, 344–56.

145. **Xiang, J., C. Li, H. Li, R. Cao, B. Wang, X. Han,** and **J. Chen** (2015). The detection of epileptic seizure signals based on fuzzy entropy. *J Neurosci Methods*, **30**(243), 18–25.

146. **Xu, G., J. Wang, Q. Zhang, S. Zhang,** and **J. Zhu** (2007). A spike detection method in EEG based on improved morphological filter. *Comput Biol Med*, **37**, 1647–52.

147. **Yentes, J. M., N. Hunt, K. K. Schmid, J. P. Kaipust, D. McGrath,** and **N. Stergiou** (2012). The Appropriate use of Approximate Entropy and Sample Entropy with Short Data Sets. *Ann Biomed Eng*, **41**(2), 349–65.

148. **Yoo, J., L. Yan, D. El-Damak, M. Altaf, A. Shoeb,** and **A. Chandrakasan** (2012). An 8 channel scalable EEG acquisition SoC with patient-specific seizure classification and recording processor. *IEEE J Solid State Circuits*, **48**(1), 214–28.

149. **Zandi, A. S., M. Javidan, G. A. Dumont,** and **R. Tafreshi** (2010). Automated Real-Time Epileptic Seizure Detection in Scalp EEG Recordings Using an Algorithm Based on Wavelet Packet Transform. *IEEE Trans Biomed Eng*, **57**(7), 1639–51.

150. **Zhang, C., L. Tong, Y. Zeng, J. Jiang, H. Bu, B. Yan,** and **J. Li** (2015). Automatic Artifact Removal from Electroencephalogram Data Based on A Priori Artifact Information. *Biomed Res Int*, **2015**, 1–8.

151. **Zhao, W., W. Zhao, W. Wang, X. Jiang, X. Zhang, Y. Peng,** and **et al.** (2020). A Novel Deep Neural Network for Robust Detection of seizures using eeg signals. *Comput. Math. Methods Med.*, **2020**.

152. **Zhou, M., C. Tian, R. Cao, B. Wang, Y. Niu, T. Hu, H. Guo,** and **J. Xiang** (2018). Epileptic Seizure Detection Based on EEG Signals and CNN. *Front. Neuroinform.*

153. **Zifkin, B. G.** and **G. Avanzini** (2009). Clinical neurophysiology with special reference to the electroencephalogram. *Epilepsia,*, **50(Suppl.3)**, 30–38.

154. **Zweig, M.** and **G. Campbell** (1993). Receiver-operating characteristic (ROC) plots: a fundamental evaluation tool in clinical medicine. *Clin Chem*, **39**(4), 561–77.

CPSIA information can be obtained
at www.ICGtesting.com
Printed in the USA
BVHW051430070423
661958BV00011B/355